AF538627

BASICS OF AGRICULTURAL CHEMISTRY

BASICS OF AGRICULTURAL CHEMISTRY

Dewasish Choudhary

ANMOL PUBLICATIONS PVT. LTD.
NEW DELHI - 110 002 (INDIA)

ANMOL PUBLICATIONS PVT. LTD.
H.O.: 4374/4B, Ansari Road, Daryaganj,
New Delhi-110 002 (India)
Ph.: 23278000, 23261597
B.O.: No. 1015, Ist Main Road, BSK IIIrd Stage
IIIrd Phase, IIIrd Block,
Bangalore - 560 085 (India)
Visit us at: www.anmolpublications.com

Basics of Agricultural Chemistry

First Edition, 2009
ISBN 978-81-261-3769-5

PRINTED IN INDIA

Printed at Balaji Offset Press, Delhi.

Contents

Preface

Chemical analyses of Soil, Plant material, Fertilizer, Irrigation water, etc. are a part of basic work in Agricultural Chemistry. The main aim of this book is to facilitate the students in the field of Soils, Plants, Irrigation, Water and Fertilizers etc. It has been compiled for the practicing chemists in research laboratories and students.

In case of soils, methods described are for nutrients considered as immediately available for providing an assessment of the available nutrient status. These plant analyses are valuable for students, researchers and plant analysis laboratories for advisory purposes. This book has been prepared to meet the needs of Teaching, Research, Fertilizer production, Advisory services, Soil, Plant, and Water testing laboratories. This book will be useful for Educational Institutions interested in Soil science, Plant analysis, water management, and fertilizers.

Author

Chapter 1

Introduction

Agricultural chemistry must be considered within the context of the soil ecosystem in which living and nonliving components interact in complicated cycles that are critical to all living things. Carbon inputs from photosynthetic organisms ultimately provide the fuel for many soil organisms to grow and reproduce. Soil organisms, in turn, promote organic carbon degradation and catalyze the release of nutrients required for plant growth.

The stability and productivity of agricultural ecosystems rely on efficient functioning of these and other processes, whereby carbon and nutrients such as nitrogen and phosphorus are recycled. Human-induced perturbations to the system, such as those that occur with pesticide or fertilizer application, alter ecosystem processes, sometimes with negative environmental consequences.

INORGANIC COMPONENTS OF THE AGRICULTURAL ECOSYSTEM

Soil is the primary medium in which biological activity and chemical reactions occur. It is a three-phase system consisting of solid, liquid, and gas. Approximately 50 percent of the volume in a typical agricultural soil is solid material classified chemically as either organic or inorganic compounds. Organic materials usually constitute 1 to 5 percent of the weight of the solid phase. The remainder of the soil volume is pore space that is either filled with gases such as CO_2 and O_2, or water.

Surface area and charge characteristics of the inorganic

portion of the solid phase control chemical reactivity. Soil particles are classified based on their size, with sand-sized particles having diameters of 2 to 0.05 millimeters (0.08 to 0.002 inches) and silt-sized particles from 0.05 to 0.002 millimeters (0.002 to 0.00008 inches). Clay-sized materials of less than 0.002 millimeters (0.00008 inches) in diameter have the largest surface area per unit weight, reaching as much as 800 meters (2,625 feet) squared per gram. Because of large surface areas, clay-sized materials greatly influence the sorption of chemicals such as fertilizers and pesticides and play a major role in catalyzing reactions.

Crystalline layer silicates or phyllosilicates present in the clay-sized fraction are especially important because they function as ion exchangers. Most phyllosilicates have a net negative charge and thus attract cations. This cation exchange capacity (CEC) controls whether plant nutrients, pesticides, and other charged molecules are retained in soil or if they are transported out of the soil system. In contrast, aluminum and iron oxides also present in the clay-sized fraction typically possess a net positive charge or an anion exchange capacity (AEC). Soils in temperate regions are dominated most often by solid phase materials that impart a net CEC, whereas soils in tropical regions often contain oxides that contribute substantial AEC.

ORGANIC COMPONENTS OF THE AGRICULTURAL ECOSYSTEM

Organic materials contained within the solid phase, although only a small percentage of the total soil weight, are extremely important in controlling chemical and physical processes in soil. Organic matter exists in the form of recognizable molecules such as proteins and organic acids, and in large polymers called humic materials or humus. Humus is dominated by acidic functional groups ("OH and "COOH) capable of developing a negative charge and contributing substantial CEC. These large polymers possess a three-dimensional conformation that creates hydrophobic regions important in retaining nonionic synthetic organic compounds such as pesticides. Nonionic pesticides partition into these hydrophobic regions, thereby decreasing off-site movement and biological availability.

A wide variety of organisms live in soil, including microorganisms not visible to the naked eye such as bacteria, fungi, protozoa, some algae, and viruses. Bacteria are present in the largest numbers, but fungi produce more biomass per unit weight of soil than any other group of microorganisms. Much of agricultural chemistry as it relates to nutrient cycles, pesticide transformation, plant growth, and organic matter degradation involves the participation of microorganisms. Microorganisms produce both intracellular and extracellular enzymes that increase reaction rates, oxidize and reduce organic and inorganic compounds, and synthesize organic molecules that modify soil chemical and physical properties.

Additional organisms in soil such as insects, nematodes, and earthworms also alter the soil ecosystem in a manner that directly or indirectly affects chemical reactions. These organisms physically process plant-derived organic materials prior to biochemical degradation by microorganisms. Nutrient release from organic materials is thus accelerated because the meso- and macrofauna expose more organic matter surface area to microbial breakdown and redistribute such materials in soil to areas of intense microbial activity.

In addition, bioturbation may also cause physical changes to the soil structure that increase pore space or modify water movement. Changes in O_2 concentration or soil water content will control biotic and abiotic reactions, altering rates of nutrient cycling and organic matter degradation.

Plant roots also modify soil by producing a zone of intense biological activity called the rhizosphere. This is a region of soil influenced by the root, most often delineated by comparing microbial numbers at greater distance from the root surface. Carbon compounds exuded or sloughed off from roots are used as a food source by microorganisms, thereby causing increased growth and activity. Microbial numbers above those of the bulk soil, which displays no root influence, indicate that the rhizosphere extends to 5 millimeters (0.2 inches) or less.

Rhizosphere microorganisms that capitalize on carbon from the plant root interact physically and biochemically with the root, potentially producing positive or negative effects on plant growth.

SOIL CHEMISTRY

Biological availabilities and transport phenomena of ions and molecules in soil are controlled by the type of bonding that occurs with the solid phase. Ions such as those typically formed when amending soils with inorganic fertilizers interact with high surface area clay and humic colloids to form either outer- or inner-sphere complexes. Outer-sphere complexes result when ions, electrostatically attracted to an oppositely charged colloidal surface, retain their shell of hydrating water molecules.

These loosely held ions satisfy the excess positive or negative charge of the colloid, but are separated from the colloid's surface by one or more layers of water. In contrast, inner-sphere complexes form when the ion loses its hydration water to form a much stronger covalent bond with the colloid. Nutrient ions held in outer-sphere complexes are plant-available because they may be exchanged with ions of the same charge, but nutrients held by an inner-sphere mechanism are not available until the covalent bond is broken.

Most soils contain a net CEC often reported in centimoles of charge per kilogram of soil (cmol/kg). Biological and physical characteristics of the soil are controlled by the amount of CEC and the specific cations involved. Soils dominated by high surface area clays or humus display the highest CECs, whereas soils with large amounts of sand or silt, and only small amounts of humus, exhibit much lower CECs.

Highly charged cations with small hydrated radii such as Al^{3+} are more tightly held on the CEC and less likely to exchange than larger, less highly charged cations such as Na^+. This general relationship is superseded when a specific inner-sphere complex forms such as between Cu^{2+} and humus, or K^+ and clay. An even more dramatic example is that of two plant nutrients, NO_3^+ and PO_4^{3+}. Negatively charged NO_3^+

readily leaches out of soil, but PO_4^{3+} is retained quite strongly because it forms an inner-sphere complex.

The percentage of the CEC occupied by specific cations influences soil pH and associated characteristics relevant to plant growth and soil biological activity. Only the most strongly held cations remain in soils in high rainfall areas. Al^{3+} dominates the CEC, hydrolyzing when released from the solid phase to the soil solution to form acidic soils with pH values often below 5.

$$Al^{3+} + H_2O \rightarrow AlOH^{2+} + H^+$$

In contrast, soils located in lower rainfall areas accumulate less tightly bound cations such as Ca^{2+}, Mg^{2+}, K^+, and Na^+ and have higher pH values between 5 and 7. In the most arid regions, large amounts of OH^--generating sodium and calcium salts accumulate, causing soil pH values to exceed 7. Plant growth is optimal in soils having pH values between 5.5 and 6.5 because aluminum toxicity occurring at lower pH values, and nutrient limitations caused by higher pH values, are avoided.

SOIL MICROBIOLOGY AND BIOCHEMISTRY

Biochemical transformations catalyzed largely by microorganisms are required for the sustained productivity of all ecosystems. Nutrients sequestered in organic materials and added in the form of fertilizers are cycled by microorganisms in their quest for energy, reducing equivalents, and carbon. Microorganisms grow and reproduce by oxidizing organic or inorganic materials, thereby releasing electrons. The electrons are passed down a series of carriers aligned in a thermodynamic gradient designed to capture energy in the form of adenosine triphosphate (ATP).

Additional electrons originating from organic or inorganic materials are used to provide reducing equivalents necessary for synthesizing cell constituents. Carbon for cell growth is obtained from the organic materials being oxidized or captured in the form of CO_2 if inorganic materials are being oxidized.

Reduction-oxidation processes are therefore central to

agricultural chemistry because oxidation of the electron source and reduction of the electron sink profoundly modify the respective element's chemical characteristics, and thus its behaviour and biological availability in the environment. For example, microbial oxidation processes convert organic compounds to CO_2, a gas, and NH_4^+, a cation, to NO_3^-, an anion. Electrons obtained in these oxidations are passed on to a terminal electron acceptor. Microorganisms use terminal electron acceptors in a sequence that maximizes energy yield starting with O_2 and proceeding through NO_3^-, Mn^{4+}, Fe^{3+}, SO_4^{2-}, and finally CO_2, which upon reduction yield H_2O, N_2, Mn^{2+}, Fe^{2+}, H_2S, and CH_4, respectively.

HUMAN MANIPULATION OF AGRICULTURAL ECOSYSTEMS

Food and fibre production are typically optimized by carefully managing the agricultural ecosystem. Synthetic organic compounds are often applied

to control plant pests including weeds, insects, nematodes, and fungal pathogens. Pesticide fate is controlled by sorption to the solid phase and degradation rate. Because most soils have a CEC, cationic pesticides are so strongly held by soil that they are typically biologically unavailable. Weak acid pesticides containing carboxyl, phenolic hydroxyls, or aminosulfonyl functional groups are weakly retained by soil and thus most likely to leach or move off-site. Weak bases, which may exist as positively charged or uncharged molecules, and nonionic compounds, are intermediate in their susceptibility to move off-site and cause environmental contamination. However, rapid degradation of some pesticides to form benign products eliminates the time available for transport, decreasing the potential for environmental problems. Both biotic and abiotic mechanisms catalyze degradative reactions, the rate of which is controlled by the pesticide's chemical structure.

With the advent of molecular techniques and the ability to transfer genes, an additional area of concern has emerged the introduction of foreign genes into plant species for enhanced crop productivity. In addition, we have the ability to produce a variety of pharmaceutical chemicals in genetically modified plants using what has been termed "pharm crops." David Suzuki and Holly Dressel in *From Naked Ape to Superspecies* have commented on such genetic manipulations, addressing the risks of placing genes from one species into another. Not only is direct gene transfer from one living organism to another possible, but extracellular DNA preserved in soil systems is also potentially available for transfer, further increasing environmental risks.

Agricultural chemistry is most often linked to food and fibre production, specifically for human consumption. Jared Diamond in *Guns, Germs, and Steel* argues quite convincingly that it was our ability to domesticate crops and eliminate the need for hunting and gathering that allowed for the establishment of permanent settlements and the development of technologically advanced societies.

The ensuing increase in human population has led to tremendous pressure to produce additional food from finite resources. Increased agricultural production, in combination with additional resource consumption and waste generation, has caused environmental degradation. By understanding key concepts in agricultural chemistry, we can utilize the soil resource to produce an adequate food supply and protect the environment.

MODERN CHEMISTRY

Modern agriculture depends quite heavily on the advances that have been made in science and chemistry in particular, to maximize the yield of crops and animal products. Fertilizers, pesticides, and antibiotics play ever increasing roles in this field. Fertilizers are perhaps the most widely used form of chemical in agriculture. Fertilizers are added to the soil in which crops are growing to provide nutrients required by the plants. Fertilizers can be divided into two categories organic and inorganic. Organic fertilizers are derived from living systems and include animal manure, guano (bird or bat excrement), fish and bone meal, and compost.

These organic fertilizers are decomposed by microorganisms in the soil to release their nutrients. These nutrients are then taken up by the plants. Inorganic or chemical fertilizers are less chemically complex and usually more highly concentrated. They can be formulated to provide the correct balance of nutrients for the specific crop that is being grown. Both organic and inorganic fertilizers supply the nutrients required for maximum growth of the crop. Inorganic fertilizers contain higher concentrations of chemicals that may be in short supply in the soil. The major or macro- nutrients in inorganic fertilizers are nitrogen, phosphorous, and potassium. These fertilizers also may provide other nutrients in much smaller quantities (micro-nutrients).

With the expansion of cities due to increases in population, there has been a loss of agricultural land. Appropriate use of fertilizers to increase crop yield has in part counterbalanced this loss of land. The use of fertilizers is not without controversy, however. There are concerns that adding supplements of nitrogen, particularly in the form of inorganic fertilizers, can be detrimental. It is thought that adding additional nitrogen to the soil can disrupt the action of nitrogen-fixing bacteria, an important part of the nitrogen cycle. If these nitrogen-fixing bacteria in the soil are killed, then less nitrogen is added naturally. As a consequence, more and more fertilizer must be applied.

Inorganic nitrogen fertilizers are relatively cheap and they are often added to arable land in excessive amounts. The crop does not assimilate all of this extra nitrogen, but instead the nitrogen can run off the land and enter the water supply. High levels of nitrogen in water can lead to eutrophication which can trigger algal and bacterial blooms. These organisms remove oxygen from the water faster than it is replaced by diffusion and photosynthesis, causing some other aquatic animals to die from oxygen deprivation. High levels of nitrate in drinking water, which can be due to agricultural runoff, have been implicated in human health problems, such as blue baby syndrome (methemoglobinemia).

Pesticides are another important group of agricultural

chemicals. They are used to kill any undesired organism interfering with agricultural production. Pesticides can be divided into fungicides, herbicides, and insecticides. Fungicides are used to control infestations of fungi, and they are generally made from sulfur compounds or heavy metal compounds. Fungicides are used primarily to control the growth of fungi on seeds. They are also used on mature crops, although fungal infestations are harder to control at this later stage.

Herbicides are weed killers that are used to destroy unwanted plants. Generally herbicides are very selective, since they would be useless for most applications if they were not. A general non-selective herbicide can be used to clear all plants from a particular area. However, appropriate treatment must be carried out to remove the herbicide or render it ineffective if that area is to be used for subsequent plant growth. Herbicides can be used to kill weeds that grow among crops and reduce the value of the harvest. They can also be used to kill plants that grow in fields used for grazing by animals, since some plants can be poisonous to livestock or can add unpleasant flavours to the meat or milk obtained from the livestock.

Breeding and genetic manipulation are used to introduce herbicide resistance to crops, allowing the use of more broad-spectrum herbicides that can kill more weed species with a single application. Herbicides include a wide range of compounds, such as common salt, sulfates, and ammonium and potassium salts. In the 1940s 2, 4-D (2, 4 trichlorophenoxyacetic acid) was developed and this herbicide is still widely used today. The use of a related compound, 2, 4, 5-T (2, 4, 5 trichlorophenoxyacetic acid), is now controlled because of its potentially harmful effects. 2, 4, 5-T was a constituent of Agent Orange, a defoliant used during the Vietnam War.

Insecticides are chemicals that are used to kill insect pests. Insects can spread livestock diseases, can eat stored grain, and can feed on growing crops. Not all insects are harmful, and certain species of insects are needed to pollinate plants to ensure that they set seed. Many insecticides are non-selective and kill all insects, beneficial as well as harmful. Some insecticides, which are very effective at killing insects, have other problems associated with them. For example, DDT (dichlorodiphenyltrichloroethane) persists in the environment and is concentrated in the food chain.

With high levels of exposure, DDT can directly kill fish and birds by paralyzing their nerve centres. In lower concentrations, it can weaken bird's egg shells and cause sharp declines in reproductive rates. Insecticides work in a number of ways. Some are direct poisons (chrysanthemic acids, contact poisons, systemic poisons), while others are attractants or repellents that move the insects to a different location (fumigation acrylonitrile). Some insecticides will only attack a particular stage of an insect's life cycle and this can make them more specific.

Antibiotics and growth hormones are routinely used as feed supplements for a number of animals. These additives are supplied to keep the animals free from disease and to help them grow to a marketable size as quickly as possible. However, the indiscriminate use of antibiotics can cause problems, since this can lead to the development of resistant strains of microorganisms or sensitization to the antibiotic among people who eat these animal products. The effects on humans of eating the meat of animals treated with growth hormones are poorly understood at the present time.

Agricultural chemistry has provided us with more and cheaper food than ever before. It has also allowed food to be produced in areas that previously were unsuitable for agriculture. The application of chemicals to farming has been one of the chemical success stories of the twentieth century. This is not to say that there have not been problems, the most famous being DDT. In the 1980s and 1990s there has been a backlash against the application of chemicals to foodstuffs in the western world. This has led to the production of organic and green products that are produced without artificial application of chemicals. These products are often more expensive to produce and this increased price is passed on to the consumer. However, there is much research to suggest that appropriate organic techniques can be competitive in cost with typical chemical agriculture and that Third World countries would benefit greatly from many organic practices.

In addition to the applications outlined above, chemicals also have other agricultural uses. For example, sulfur dioxide can be used to keep grain fresh and useable for a longer period of time than untreated grain. Other chemicals can be added to promote the ripening of fruits or the germination of seeds. It is difficult to estimate the monetary value of agricultural chemicals, but many multi- national corporations are involved in their manufacture and use. Agricultural chemistry has increased the diversity of the human diet and has led to a greater overall availability of food, both animal and plant.

Chapter 2

Soil Development

Soil-Parent-Material

In the late eighteenth and early nineteenth centuries many farmers and amateur geologists became interested in the reasons for the distribution of the different types of soil. They came to the general conclusion that there was a direct and fairly simple relationship between the character of the soil and the nature of the rock beneath it. They found chalky soils on chalk, sandy soils on sandstone, clayey soils on shales and organic soils on peat and so they devised soil classifications which were based on the nature of the underlying soil-parent-material.

Nearly all these workers had experience of only small areas however; it was not until the latter part of the nineteenth century that individual pedologists gained a knowledge of soils on a continental scale. When this was achieved there ensued a change in attitude towards soil classification. Soils of very similar character were found in great belts extending across large parts of European Russia and the U.S.A. and individual soil belts frequently crossed rock outcrops of very diverse lithology. It became apparent that other factors, particularly climate and vegetation, were just as important as parent-material in determining the ultimate nature of the soil.

Weathering

The fact remains, however, that much of the material composing most soils has been derived from the underlying rock; no matter how long a soil has been in process of formation, the original 'raw material' must therefore have some effect on its ultimate nature. The process by which the parent-rock-material is broken into small fragments are known collectively as *'weathering'*.

Two distinct types of weathering occur. Firstly there are those by which rock is broken down into progressively smaller pieces without any change in the chemical composition of the component minerals. The shattering of quite hard rocks by frost action and the splitting up of fissile, argillaceous rocks like shales, by alternate wetting and drying, are good examples of this and they are known collectively as *mechanical weathering*. Secondly there are the processes of *chemical weathering.*

These not only cause the original rock to be split into smaller fragments; they also bring about chemical alterations in some of the minerals composing the rock. Many rock minerals are attacked by such substances as dissolved carbon dioxide (CO_2) derived from the atmosphere and by organic acids from decaying vegetation.

Both types of weathering usually operate together on the same site but usually one of them exceeds the other one in importance. Thus, in cold and dry regions, or in places where the overlying soil is thin or nonexistent, mechanical weathering is the more important. On the other hand, in hot and wet regions, or in places where a thick soil overlies the rock, chemical weathering predominates.

The Products of Weathering

Rocks vary enormously in their physical and chemical properties; similarly the weathered materials derived from different rocks are themselves very different. The physical and chemical properties of individual *weathering complexes* are so complex that simplified statements can be very misleading; nevertheless a number of valid generalisations can be made, to show how soils develop and function. From the point of view of the soil scientist it is probably most useful to consider weathering products in two categories; on the one hand there are those which are readily soluble and on the other those which are almost insoluble. The more soluble ones are predominantly carbonates since carbonic acid ($H_2 CO_3$) occurs almost universally in soils. The atmosphere contains small amounts of carbon dioxide (CO_2) and any raindrop inevitably dissolves a little of this as it falls to earth

$$CO_2 + H_2O = H_2CO_3$$

This acid, though chemically weak, is able to detach substances such as potassium, magnesium and calcium from rock minerals and to bring them into the soil solution as carbonates (K_2CO_3, $MGCO_3$, $Ca(HCO_3)_2$). These bases contain some of the most important mineral foods required by plants. The almost insoluble products of chemical weathering comprise a vast number of chemical substances amongst which, however, only a few types are common. These poorly-soluble substances go to form what is usually called the *inorganic fraction* of the soil.

As an illustration one may select the weathering complex which develops on a granite surface. The simplest type of granite consists almost entirely of

- Orthoclase felspar (potassium aluminium silicate $K_2O.Al_2O_3.6SiO_2$)
- Biotite mica (a mixed silicate of potassium, magnesium, iron and aluminium $K_2Mg_6O.2Fe_2O_3.Al_2O_3.6SiO_2.2H_2O$)
- Quartz (silica SiO_2)

On being weathered chemically, the felspar loses its potassium in the form of potassium carbonate (K_2CO_3) and a residue of a fairly simple clay mineral called kaolinite ($Al_2O_3.2SiO_2.2H_2O$) is left. The biotite mica breaks down to form potassium and magnesium carbonates (K_2CO_3, $MgCO_3$), the sesquioxides of iron and aluminium (Fe_2O_3, Al_2O_3) along with some clay (*vide infra*). The quartz, which is very inert chemically, is merely released from the rock as sand grains when the other minerals disintegrate. The clay, sand and the sesquioxides released from the minerals are added to the inorganic fraction of the soil while the soluble carbonates enter the soil solution.

Similar kinds of change take place when other types of rock are weathered although usually the nature and variety of the substances produced is even more complex. For instance, the clay minerals produced by the weathering of many rock minerals have a much more complicated structure than kaolinite; most clay minerals are mixed silicates of iron and aluminium (*vide infra*). Many of the minerals of igneous and metamorphic rocks also contain a much greater variety of bases to be liberated as carbonates. Furthermore, whereas acid igneous rocks like granite contain a great deal of quartz and thus weather to a predominantly sandy material, basic igneous rocks such as basalt produce a residue which is predominantly clay.

Generally speaking, a sedimentary rock produces a weathered residue which is much more homogeneous than that from an igneous rock. This is because sedimentary rocks are composed of material which has, to a greater or lesser extent, been sorted during a previous cycle of erosion and deposition. Shales thus weather to a predominantly clayey or *argillaceous* residue while sandstones and grits produce a sandy or *arenaceous* one. Even then, however, different types of sandstone give rise to weathering residues of different fertility; some sandstones consist of grains cemented together with calcium carbonate ($CaCO_3$) while others are cemented by yet more quartz.

The former will obviously release a great deal of lime during weathering while the latter, as well as being much harder and more resistant to weathering, will produce remarkably little soluble material during the process. The above are merely a few examples of the lithological contrasts found in different rocks; any good textbook of physical geology will provide some account of the mineral contents of all types of igneous, metamorphic and sedimentary rocks from which one can obtain a clear indication of their inherent potentialities as soilparent-materials

From the point of view of the pedologist, differences in the lithological characteristics of the parent rock are of salient importance for two main reasons. Firstly, rocks differ enormously in the amount of soluble plant nutrients that they liberate on weathering. Secondly, the residue of weathering can vary between almost 100% clay and almost 100% quartz although most rocks yield a residue which contains a certain amount of both. This is of considerable importance even from the purely physical standpoint. Individual clay particles are exceedingly small, all having a diameter of less than.002 mm.

Quartz grains on the other hand are nearly all much larger in size, those with a diameter between.002 mm. and.02 mm. being referred to as *silt* and those with above.02 mm. as *sand*. It is clear that the larger the percentage of clay particles in weathered material, the slower will be the rate at which water will percolate through it. On the other hand, very sandy materials will allow water to flow through them almost unimpeded. Soil textures are classified as sandy, silty, clayey or loamy according to the proportions of the different types of particle present.

Apart from the purely physical property of permeability, however, the properties of soil chemistry and fertility are also very closely bound up with the relative amounts of clay and quartz in the inorganic fraction of the soil (*vide infra*).

Organic Matter in the Soil

It has already been pointed out that soil development is normally accompanied by the development of vegetation in the form of some recognisable prisere. Even as rock waste begins to be produced by weathering, some organic material also accumulates as the roots wither and aerial organs of the plants die down. The effect of organic material must therefore be taken into account from the very first stages of *pedogenesis*. In mature soils, because of differences in rates of growth and decay, the organic fraction may vary between almost 100% (as in peat) and much less than 1% (as in many desert soils).

Organic matter which has been partially decomposed and incorporated into the soil is usually spoken of as *humus*. When plant organs die they initially suffer a certain amount of decay and disintegration; generally speaking this takes place very quickly in the tropics and it is also quite rapid in higher latitudes in summer. Everywhere, however, the rate of decay appears to slow down markedly when a particular stage of disintegration has been reached. It seems as though some of the lignocellulose or woody material of the original vegetable matter combines with certain compounds of the plant protein to form substances which are much more resistant to the attacks of decay organisms.

Although the details of their molecular formulae have not yet been ascertained, these decay-resistant substances probably comprise a series of distinct chemical compounds; they are not merely undifferentiated masses of organic material. Nevertheless there are numerous grades of humus, the type varying according to the nature of the associated plant community. Some communities are composed of plants which demand relatively large quantities of mineral nutrients from the soil; they abstract substances like potassium, calcium and ammonium (K, Ca, NH_4) and incorporate them in their leaves, stems and roots.

Consequently these organs are rich in mineral matter and nitrogen. When the plant organs die the nutrients are returned to the soil where they tend to be retained by the humus. Such humus has, in consequence, an alkaline reaction and is known as *'mild humus'* or *'mull'*. On the other hand many plant communities are composed of plants which are adapted to an environment which is very poor in nutrients. They require little in this respect and therefore they have little to return. The humus they yield is thus poor in minerals and acid in reaction and is referred to as *'raw humus'* or *'mor'*. Although it is customary to divide humus into these two categories, it must be realised that all gradations between very alkaline mull and the most acid mor can be found.

Humus produced by the disintegration of the fibrous rooting systems of herbaceous plants is already in a finely-divided state and is disseminated throughout the soil. On the other hand the dead stems and leaves of plants fall on the surface of the soil where they humify to form a distinct layer of peat, turf or leaf-mould. This organic layer frequently becomes quite thick, particularly when it is composed of mor or in places where the ground is persistently waterlogged. Acidity and wetness tend to preserve a surface layer of peat or mor for two reasons; firstly, decay bacteria are much reduced, both in numbers and activity, by both these conditions; secondly, the larger members of the soil fauna such as earthworms and beetles are either few in number or absent altogether in both acid and waterlogged soils.

In more basic and well-drained soils the opposite is usually the case; decay bacteria operate quickly in the destruction of humus while earthworms and other organisms mix it with the underlying inorganic material, often with astonishing rapidity. In his treatise on the earthworm Darwin demonstrated that these creatures can deposit the equivalent of ten inches of top soil in the form of worm-casts on the surface of a pasture in the course of only a few decades. These casts are composed in great part of inorganic material derived from below. This becomes mixed with dead organic material which would otherwise accumulate on the surface.

Soil Structure

The incorporation of humus into the inorganic, weathering material of the soil does not result merely in a simple mixture of organic and inorganic particles. There is abundant evidence that the humus compounds enter into quite intimate chemical linkages with the tiny clay particles to form what is usually referred to as the '*clay-humus complex*'. In fact, from the point of view of soil structure and soil fertility, clay particles resemble humus molecules much more than they resemble the coarser fragments of silt and sand.

The latter are usually referred to as the '*soil skeleton*' and are quite inert chemically. On the other hand, the clay-humus complex is chemically active; not only do the large molecules of clay and humus effect chemical linkages between themselves, they also have many spare *valencies* which they use to retain mineral elements in the soil. If this active complex were not present in the soil, one substantial shower of rain would be sufficient to wash out all the highly soluble salts of potassium, magnesium and ammonium. Inorganic salts, when dissolved in water, *ionize* to a considerable extent. Dissolved molecules of potassium sulphate in the soil solution are thus partly dissociated into separate potassium and sulphate ions

Each of the free potassium ions carries a positive charge or valence with which it is able to re-link with the sulphate ion or with any other similar ion in the soil. In fact the large molecules of the clay-humus complex are chemically analagous to the sulphate ion so that the potassium ions can effect loose, temporary linkages with these soil compounds. Magnesium, calcium, ammonium and hydrogen ions, along with many others, can be retained similarly by the clay-humus. The actual molecular structures so formed are very complex but it is quite permissible to represent them diagrammatically.

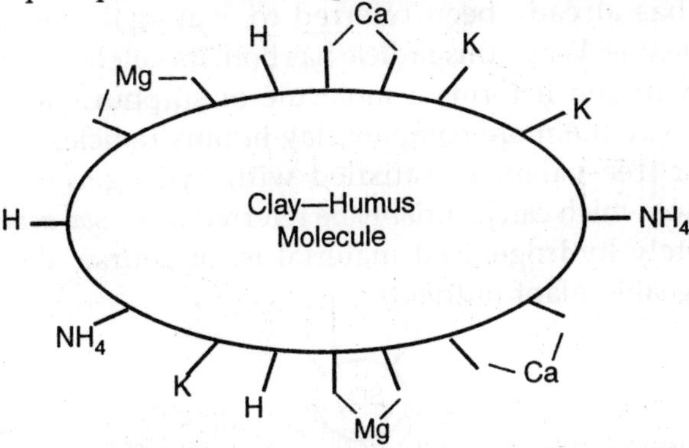

Fig. Diagrammatic representation of a clay-humus Molecule with Appurtenant ions of Hydrogen and Nutrient Minerals.

It is clear, therefore, why a mere mixture of silt and sand, completely devoid of clay and humus, could never be made into a fertile soil. Vast amounts of soluble mineral fertiliser could be applied to such a coarse, skeletal mixture but would be lost entirely after only a few hours' rain. On the other hand, soils with a large clay-humus fraction have the potential for holding plant nutrients in chemical bonds for considerable periods. Such nutrients can be detached by the root hairs of plants, how ever, as they push their way between the soil particles.

In order to understand how soil development takes place under natural conditions and thus to appreciate how soils like podzols and chernozems come into existence, it is necessary to explore the chemistry of soils a little further. It would be easy to obtain a misleading impression from the points that have just been made. In spite of the fact that the clay-humus molecules can retain mineral ions temporarily, the latter are constantly being detached by the purely inorganic action of rain-water. Because of thisv, soils in humid regions soon become completely devoid of exchangeable mineral ions unless the latter are constantly being replaced.

When the clay-humus molecules lose their nutrient ions because of percolating rain-water, increased acidity must occur because hydrogen ions are substituted. The sulphate radicle, which has already been referred to, can again be used to illustrate this. When this radicle has both its valencies satisfied with hydrogen it forms a molecule of sulphuric acid. In a similar way the more complex clay-humus radicles can have all their free valencies satisfied with hydrogen to form a substance which can justifiably be referred to as *'soil acid'*. Such completely hydrogenised material is, of course, devoid of exchangeable plant nutrients.

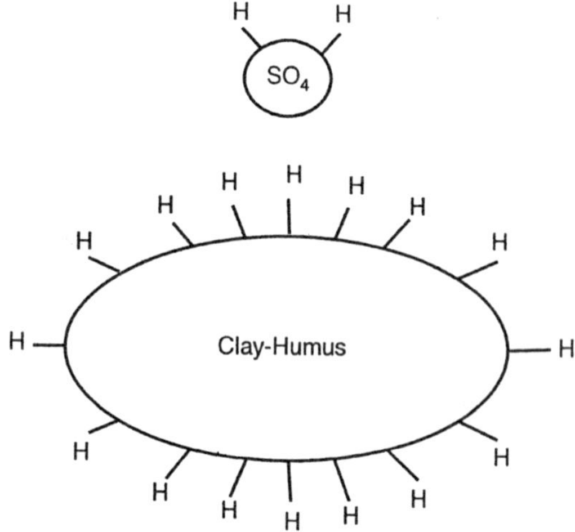

Fig. Diagrammatic Representation of Sulphuric Acid and Clay-humus Acid Molecules.

Leaching

It is important to recollect that rain-water is, in effect, a dilute solution of carbonic acid (H_2CO_3). Consequently, in a region of humid climate where downward percolation of rain-water is persistent, the nutrient ions which are detached are carried away into the drainage water in the form of carbonates. The carbonic acid molecules exchange their hydrogen for mineral ions as they pass through the clay-humus complex and then continue to move downwards as carbonate salts. It is clear, therefore, that unless replacement takes place, either through the further weathering of rock minerals or by decay of organic matter, ultimately the clay-humus complex will be completely hydrogenised and become pure soil acid. The process of down-washing is known as *leaching*.

Base Status and pH value

The actual content of exchangeable mineral ions is referred to as the base status of a soil. In theory the concept of base status is easy to appreciate but, in practice, its determination is difficult. One method of obtaining some index of the state of the clay-humus complex is by measuring the acidity of the soil solution. The assumption is that, the more exchangeable hydrogen there is attached to the clay-humus radicles, the less 'room' there will be for mineral ions.

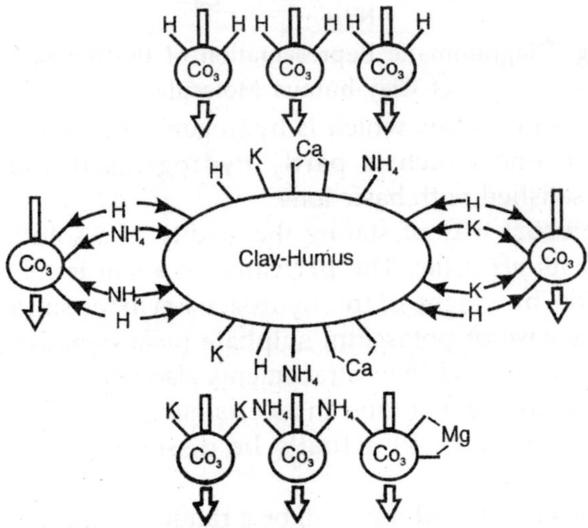

Fig. Diagrammatic Representation of the Leaching of Nutrient Ions by Percolating Carbonic Acid.

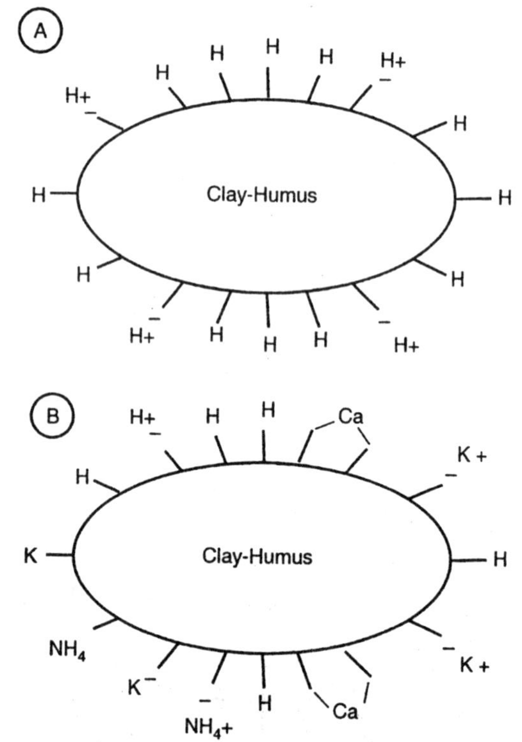

Fig. Diagrammatic Representation of the Ionization of Clay-humus Molecules

- In a complex which is hydrogen saturated.
- In one which is partly hydrogenised and partly satisfied with basic ions.

The usual way of stating the acidity of a solution is in terms of its *pH value*. The pH value of a soil is merely the proportion, by weight, of free hydrogen ion in the soil solution. As in the case of potassium sulphate (*vide supra*), the clay-humus radicles and their attachments also ionise to a certain extent. At any point of time an ascertainable percentage of the appurtenant ions will actually be dissociated in the soil solution.

These dissociated ions will be a random sample of all the ions present. It follows from this that the greater the total quantity of hydrogen ion, the more ionised hydrogen will be present. In the case of a completely hydrogenised clay-humus complex therefore (A), all the dissociated ions will be hydrogen; in the case of one which is partially satisfied with hydrogen and partly with bases (B), there will be less ionised hydrogen while, in the case of a completely basesaturated complex, all the freed ions will be bases.

A solution which, by weight, is exactly 1/10,000th part hydrogen ion is said to have pH 1; one which is 1/100,000th part hydrogen ion–pH 2; 1/1,000,000th part–pH 3 and so on. It will be seen that the pH progression is a logarithmic one but that the lower the acidity happens to be, the higher the pH value. In fact the pH value of a solution is *the logarithm of the reciprocal of the hydrogen ion concentration.*

It can be demonstrated that a soil whose clay-humus complex is exactly satisfied with hydrogen, with no other acids present, will have approximately pH 4. On the other hand, one which is exactly base-saturated often has a pH of approximately 7. The latter is thus the pH value which corresponds to neutrality–neither acid nor basic. All gradations between pH 4 and pH 7 can be found, corresponding with the gradation between complete hydrogenisation and complete base-saturation of the clay-humus complex.

There are also some soils which, apart from having their clayhumus complex completely hydrogenised, also contain certain amounts of stronger organic acids (e.g. oxalic acid). These soils may have pH values of 3, 2 or even less. At the opposite extreme, most soils in sub-humid and semi-arid regions, apart from having a completely base-saturated clay-humus complex, also contain an excess of bases like calcium carbonate ($CaCO_3$). They therefore have an alkaline reaction and pH values of 8 or 9 are normal.

A statement of the pH value of a soil thus gives some indication of its fertility and one can be fairly confident that, if the pH of *a particular* soil falls, this is due to an increase in exchangeable hydrogen with a concurrent decrease in exchangeable bases. Unfortunately, however, the same kind of inference cannot be made when comparing the pH values of *different* soils. Two different soils may have similar pH values and yet contain very different amounts of exchangeable bases; conversely, two other soils may have very different pH values and yet contain very similar amounts of exchangeable bases. This can be understood by a consideration of two hypothetical cases. The first of these is a soil whose clay-humus complex is exactly base-saturated and which, in consequence, has a pH of 7. The second is a soil of quite different constitution whose clay-humus complex is approximately half-satisfied with bases and half with hydrogen and which, in consequence, has a pH of between 5 and 6 (i.e. half-way between 4 and 7).

At first glance one might infer from this information that the first of the two soils must inevitably contain a larger quantity of exchangeable bases than the second one but this is not necessarily so. If it is further postulated that the first soil consists of about 95% skeletal material and only about 5% clay and humus while the exact opposite is true of the second soil, then the two are seen in an entirely different light. Even though the first soil is base-saturated, it contains such small amounts of chemically active clay and humus that the actual *quantity* of bases being held cannot be anything but small.

On the other hand, the second soil is so rich in clay and humus that it retains much larger quantities of bases even though it is partly hydrogenated. It follows from this, that a simple statement of pH value does not, of necessity, give an accurate impression of the base status of a soil. Consequently modern soil analyses include quite separate assessments of all the main exchangeable bases such as potassium, calcium and ammonium.

Colloidal Properties

The two types of substance which combine to form the clay-humus complex have a further property in common which is important in soil development. Both clay and humus are *colloidal*. A colloid is a substance which, without actually changing its chemical composition, can exist in quite different physical states. Probably the most familiar colloid to the layman is a solution of gelatine. This substance can exist as a completely mobile fluid or as a flexible solid, the actual state, at a given concentration, being determined by temperature.

At higher temperatures it adopts the sol or fluid state and, at lower ones, the *gel* or solid state. The clay-humus colloids in the soil behave analagously though in by no means so spectacular a manner. The state, in this case, is determined not by temperature but by acidity. Soil colloids tend to be stable (gel) in a basic or neutral medium (high pH) and to become more mobile (sol) with increasing acidity (lower pH). Thus, in acid podzols, whole molecules of clay and humus become mobile and are washed downwards. The upper mineral horizons thus tend to become poorer in clay and relatively richer in skeletal material. Any humus in these uppermost mineral horizons also tends to be transient; it is in process of being moved downwards from the peaty humus of the uppermost layer.

Other changes also take place in a completely hydrogenised clay-humus complex. Not only do whole molecules become mobile and move downwards, the clay molecules also become chemically unstable and begin to dissociate. As has been noted already, most clays are complex silicates of iron and aluminium and their generalised formula can be written as follows

For practical purposes therefore they can be assumed to consist of different proportions of molecules of iron sesquioxide, aluminium sesquioxide1A sesquioxide is an oxide which has one and a half oxygen atoms to every one metallic atom. and silica though in some cases (e.g. kaolinite) the iron sesquioxide is lacking and, in others, bases such as potassium and calcium are incorporated. Generally speaking, however, when clay dissociates, the two sesquioxides and silica are the main substances to be released.

It is important to note, however, that whole molecules of clay do not disintegrate instantaneously. In some climatic conditions they shed sesquioxide molecules faster than silica molecules; this happens in the case of podzol formation. In other conditions, silica is lost faster than sesquioxides; this is the laterisation process. The clay in the upper horizons thus changes its general composition and may ultimately disintegrate altogether.

The separate molecules of sesquioxide and silica are themselves colloidal so that they too can become mobile in a soil which is periodically soaked by rain. As they move downwards along with whole molecules of clay and humus, they usually encounter different chemical conditions in the lower, less-weathered layers. Because of this they are frequently re-deposited. Thus, in most soils which are subject to leaching, it is possible to distinguish two quite distinct sets of horizons beneath the uppermost mat of organic material.

The upper set are referred to as *'eluviated horizons'* and the lower as *'illuviated'* ones. There is an almost infinite variation in the nature and arrangement of materials within this general scheme of things however. Indeed there are so many major and minor variables operating in soil development that it is quite difficult to find two soils from different sites which are alike in all their main characteristics.

Pedalfers and Pedocals

Leached and acid soils whose clay minerals have suffered some dissociation into their component parts are referred to as 'pedalfers'. Soils of this class occur, generally speaking, in areas where the amount of water reaching the surface as precipitation exceeds the amount which is evapourated. Here the excess water percolates down through the soil removing any excess bases which might otherwise accumulate.

In drier areas there is no such excess of water and here some of the less soluble bases, particularly calcium carbonate ($CaCO_3$), are able to accumulate. This ensures that the clay-humus complex remains base-saturated and furthermore, the excess of lime gives rise to a pH in excess of 7. Because of this alkalinity, the clay remains quite stable with no tendency to dissociation. These lime-accumulating soils are classed together as pedocals.

The Zonal Concept

Several main factors are operative in determining the

course of soil development under natural conditions. The importance of the lithological characteristics of the parent-material has already been demonstrated. If this is coarse and siliceous it weathers to a material which permits rapid leaching; it will also probably be poor in plant nutrients from the outset and incapable of retaining any considerable amount of nutrient which might be supplied. On the other hand, parent-material such as calcareous marl will weather to a much less pervious clay material which is rich in calcium and other nutrients from the start. From these two examples it is clear that rates of leaching and innate fertility must be influenced by the physical and chemical properties of the parent-rock.

Climate also is of obvious importance since it influences the rate at which leaching, weathering and organic decay take place. Other things being equal, leaching of nutrients and the podzolisation process will proceed much more quickly with heavy rainfall than with light rainfall. Similarly the weathering of rock minerals and the decay of organic materials go on much more quickly in a hot, moist climate than in a cool, dry one. Generally speaking, the average rate of decomposition processes, both mineral and organic, doubles with every increase of 18° F. (10° C.) above freezing-point.

The pioneers of modern soil science were so impressed with the apparent correlation between climatic and soil distributions over the continents that they devised a *zonal soil classification* which has persisted to the present day. The basic premise underlying this classification is that, regardless of the original nature of the parent-materials, given a certain set of climatic conditions, a specific type of soil will ultimately come into existence. Russian pedologists such as Glinka were the chief protagonists of this classification. They noted a belt or 'zone' of podzols across northern European Russia in a region characterised by cool, moist climatic conditions and a vegetation of coniferous forest.

Further south was a zone with warmer summers and less intense leaching where brown forest soils were found beneath a cover of deciduous forest. South again, in the Ukraine, was a zone of hotter, drier conditions in the growing season where grasslands were found prior to clearance and cultivation. Here black earths or chernozems were the commonest soil type. The zonal schools were further confirmed in their views by the discovery that soils from analagous climatic regions in North America appeared to be very similar to those of Russia. It seemed to them that climate, rather than parent-material, was the more important factor accounting for the distribution of soils on a continental scale.

Limitations of the Zonal Concept

Even at its inception, however, the zonal classification had to be hedged around with provisos. Just as all wild vegetation cannot be regarded as climatic climax vegetation, so many natural soils are obviously not 'climatic climax' or 'zonal' soils. As in the case of vegetation, there are numerous factors which arrest, deflect or in some way inhibit soil development.

Firstly, it is clearly necessary for weathered material to remain quite undisturbed in situ for a long period of time in order for weathering and pedogenic processes to produce a mature profile (i.e. one which does not alter in its essential features with further lapse of time). It is this mature soil, in equilibrium with climate, which is implicit in the concept of 'the zonal soil'. But in many places on the earth, relief and erosional forces do not permit the same material to be undisturbed, with no additions or subtractions, for long periods of time.

In any area of considerable slope the process of soil creep, even beneath dense vegetation, will be sufficiently intense to remove or mix the developing horizons long before maturity has been achieved. At the opposite extreme, on flat alluvial plains or in volcanic areas, frequent deposition of mineral-rich alluvium and volcanic ash respectively, will effectively offset the pedogenic process of leaching. Addition or removal of inorganic material thus effectively prevents the attainment of maturity; soils subject to persistent truncation or to frequent deposition are therefore referred to as *permanently immature soils*. A special category had to be made for them in the world zonal classification; they were classed together as *azonal soils*.

Apart from the latter there are other soils which are anomalous within the general zonation o soils. Firstly, there are distinct types of soil in naturally ill-drained areas. Soils with natural free drainage experience persistent percolation of water in areas of heavy rainfall; this can be prevented, however, by a high water-table. Furthermore, in areas of low rainfall, impeded drainage often prevents the leaching of the most soluble substances such as common salt (NaCl) so that they are liable to accumulate in the soil. Impeded drainage thus causes anomalous soils to develop in both humid and dry regions. Anomalous soils also develop on certain types of rock outcrop.

Particularly noteworthy are those which form on limestones in humid areas. Because limestones are composed almost entirely of calcium carbonate, the weathering complex derived from them can be particularly lime-rich. This, in many cases, is sufficient to offset losses by leaching; consequently, areas of limestone within regions of generally podzolised soils may carry soils which are actually alkaline in reaction. An island of soil thus arises which has characteristics very similar to those possessed by soils in dry regions. It is because of this that soils which, for all kinds of sporadic reasons, are anomalous within the areas where they are found, have been classed together as *intrazonal soils*.

Providing its limitations are clearly appreciated, the soil classification devised by the zonal school is a useful framework in which to view pedogenesis on the world scale. Its main limitation is comparable with that which restricts the concept of climatic climax vegetation. In both cases one factor–that of climate–is taken as the basis for classification; all other factors are regarded as subsidiary.

Thus, for an area like England and Wales where the climate is everywhere relatively humid and cool, a small-scale zonal soil map shows a homogeneous area of brown forest soils. Because England and Wales have a very varied lithology, however, the true picture of soil distribution, even under natural conditions, was very much more complicated than this. Indeed, over large areas, the zonal soil-type was the exception rather than the rule; limestone soils, truncated soils, alluvial soils and soils with impeded drainage predominated. Quite clearly, climate is not everywhere the predominant factor determining soil distribution; for areas like this one the zonal soil map may give a very misleading picture.

The zonal soil concept has other limitations. In most kinds of parentmaterial and in all types of climatic conditions, many centuries are required for the evolution of a mature soil profile. The zonal concept requires a gradual development of soil characteristics in response to climatic conditions so that, ultimately, equilibrium between soil structure and climate is reached. From then onwards, although continued weathering may cause further increase in depth, no change in the salient features of the profile will occur.

For such equilibrium to be reached it is quite clear that climatic conditions must remain constant throughout the whole period of profile evolution; if a major change in climate occurs, the whole trend of soil development will also be altered. It is well known, however, that during the past 10,000 years there have been profound climatic changes throughout much of middle and high latitudes. Many places here are experiencing very different soil-forming processes today from those which were operating in 1000 B.C. Knowledge of the climatic history of inter-tropical areas is still very scanty but here again evidence of considerable climatic change during the past 10,000 years is accumulating rapidly.

This being the case, it seems likely that many profile features which can be observed today may not be the product of present climatic conditions; they may have been formed in the past under quite different conditions and thus be mere survivals. Relict horizons of this nature may be particularly plentiful in the tropics. Some of those horizons which may traditionally have been regarded as typical features of present-day 'tropical zonal soils' may be fossil horizons of this kind.

The zonal soil concept is useful as long as these limitations are fully recognised. It cannot be denied that beneath grassland in a dry sub-humid region in mid-latitudes a zonal chernozem can confidently be expected, provided the slope is not too great and the drainage is quite free. It will have been noted, however, that many types of zonal soil appear to evolve only beneath specific types of vegetation. Indeed, it seems probable that associated vegetation may be of greater significance than climate.

The Nutrient Cycle

Many pedologists in the past, while recognising the importance of vegetation as a supplier of humus, have had an over-simplified view of the relationship between vegetation and other factors of the environment. Many protagonists of the zonal concept, while postulating that the subhumid climate was the reason for the belts of chernozem in Eurasia and North America, cannot fail to have noticed that this chernozem has only come into existence beneath grassy vegetation; it has never formed beneath forest.

They seem to have assumed, however, that the grass communities were the *inevitable* vegetation in a sub-humid climate and, this being the case, they saw no point in treating climate and vegetation as independent soil-forming factors. Many observations in the past half-century have shown this to be fallacious however. In Texas for instance, woodland and grassland are often found side by side in areas of identical climate, both types of community inhabiting soils which are freely drained. Beneath the grassland, however, a typical chernozem is found while beneath the woodland a quite distinct forest soil is maintained. A similar situation is found in the Ukraine; here the soil beneath the trees is often distinctly podzolised while that beneath the grass is an unleached chernozem.

In the light of this kind of evidence, quite clearly, vegetation must be treated as a completely independent variable; completely different types of vegetation can be found in areas of identical climate. Indeed, so specific are the effects of vegetation types on the soil profile, that many soil scientists now seem to view vegetation almost as the 'machine' in the soil-making 'factory'. It is as though the parent-rock provides much of the raw material, the climate lubricates and determines the speed of manufacture but the vegetation ultimately determines the nature of the finished product. This view probably overstates the importance of vegetation to some extent just as the earliest pedologists over-emphasised lithology and the pioneers of the zonal concept overstated the case for climate.

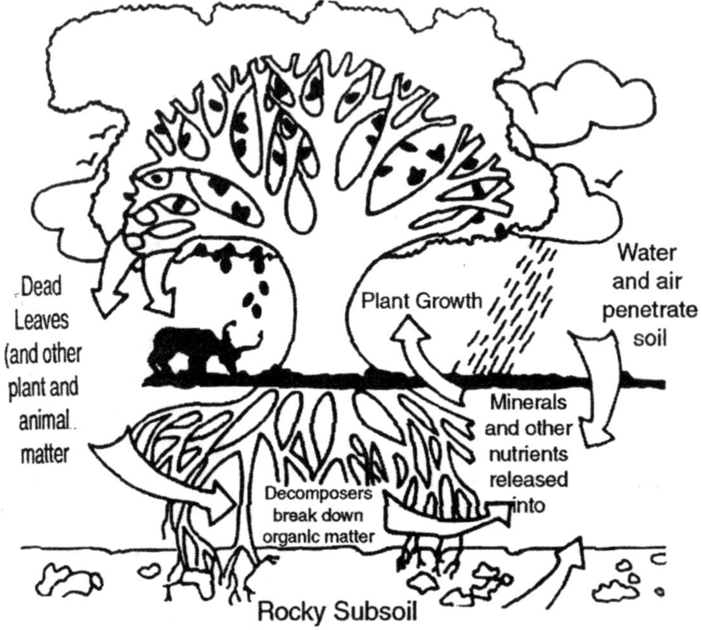

Fig. The Nutrient Cycle

Nevertheless, it cannot be denied that, without a cover of nutrient-demanding vegetation, no soil could maintain indefinitely a high base status in a humid climate, no matter how rich in bases the underlying parent-material. Without such a vegetation cover to abstract bases from the soil and to return them in base-rich humus, leaching can rapidly remove all the bases as they are released from the rock minerals by chemical weathering. There is, clearly, a nutrient cycle in which vegetation is an indispensable link.

If a less demanding vegetation type supersedes a richer one the effect may be profound; from then onwards, even though more mineral nutrient is made available by further rock weathering, soluble plant foods will be permitted to slip away rapidly and to be lost to the ground water. On the other hand, if a more demanding vegetation takes the place of a poorer one, a richer soil will gradually develop as more and more bases are incorporated into the cycle. This must have happened over large areas of lowland Britain where the original forests have given place to more demanding grass communities. Because of human interference the original brown forest soils were thus changed into soils which, in many ways, resemble true prairie soils.

Since the relative amounts of exchangeable bases and exchangeable hydrogen associated with the clay-humus complex have such an influence on the rate of processes like podzolisation and laterisation, quite clearly the richness (or poverty) of the nutrient cycle is of salient importance. The extent to which vegetation, quite independently, can affect the cycle is still open to much speculation. Some demanding plant communities may be capable of invading a poor environment and of raising the level of the nutrient cycle; other communities may be less resilient and flexible in their reactions. There can be no simple answer to the general problem; all communities and all individual environments must be studied quite independently.

NITROGEN FIXATION

Nitrogen tantalizes mankind with the paradox of poverty in the midst of plenty. All living things on this planet–animal and vegetable–must have nitrogen in their food. The earth's atmosphere contains far more than enough nitrogen to satisfy the requirements; there are some 20 million tons of it in the air over each square mile of the earth's surface. Yet the free nitrogen in the air is so difficult to incorporate into foodstuffs that man must engage in back-breaking toil to conserve the comparatively small amount that nature captures and fixes in the soil.

To be sure, it is fortunate that nitrogen is chemically inert. If it were less reluctant to combine with other elements (and its thermodynamic relations indicate that it has the potentiality of being much more active than it ordinarily is), it might readily combine with water to form nitric acid. As some authorities on thermodynamics have pointed out, "it is to be hoped that nature will not discover a catalyst for this reaction," for if it did, the oceans would turn into dilute nitric acid–a catastrophe certainly as horrible as any visualized in speculations about atomic warfare.

Nature handles nitrogen–prodding it out of its chemical sluggishness and controlling its tendency to react rapidly thereafter-by means of a complicated cycle. Certain organisms in soil and water, called "nitrogen fixers," take up the air's free nitrogen and combine it in organic compounds which are suitable for plant or animal food. Other organisms, called "nitrifiers," convert this organic nitrogen into the mineral nitrates required by plants. Still others, the "denitrifiers," decompose dead organisms and eventually return free nitrogen to the air. Some nitrogen also is lost by washing of soil and sewage into the sea.

This continual leakage of nitrogen to the atmosphere poses one of the most important problems for the survival of living forms on the earth. Man has begun to try to stop the drain by artificial methods of recapture, but his processes for combining nitrogen with other substances are expensive and succeed only in making the simplest nitrogen compounds, such as ammonium salts, nitrates, urea and cyanamide. To make up our losses and recombine nitrogen in the forms in which plants and animals need it, we are almost entirely dependent on the nitrogen-fixing organisms.

Nature's nitrogen cycle is as important to us as the carbon cycle of photosynthesis, by which plants recapture carbon dioxide from the air and convert the carbon into organic compounds. The nitrogen fixers and the photosynthetic organisms, linked in a majestic partnership, keep the living economy of the earth solvent.

What are these natural allies of man in the struggle to wrest nitrogen from the atmosphere, and how do they do it? Obviously this is a matter of considerable importance to mankind. Unfortunately we know much less about nitrogen fixation than we do about photosynthesis, which has not yet yielded all its secrets. Within the last three years, however, there has been more progress in our knowledge of nitrogen fixers than had been made in all the previous history of research on the subject.

Until 1949 it was generally believed that very few microorganisms had the ability to fix nitrogen. The best-known were the bacteria belonging to the genus *Rhizobium,* which inhabit the root nodules of leguminous plants such as peas, barley and oats. Then there were some free-living organisms, mostly lumped in a genus called *Azotobacter,* which means simply nitrogen-fixing bacteria. There were also a few known species of nitrogen-fixing bacteria that live in the absence of oxygen, for example, species of the genus *Clostridium,* some members of which are the active agents in gas gangrene and tetanus. Finally the primitive plants called blue-green algae were known to carry out nitrogen fixation under certain conditions.

Then in 1949 and immediately afterwards came a flurry of discovery which turned up undreamed numbers of other organisms that fix nitrogen. It began with a chance finding by Howard Gest and the author during a research into a subject seemingly totally unrelated to nitrogen fixation.

We were studying the phosphate metabolism of the purple bacterium *Rhodospirillum rubrum*. Rhodospirillum is one of the many genera of photosynthetic bacteria. These bacteria, like the green plants, depend on light for growth, but instead of yielding oxygen they produce carbon dioxide and other carbon compounds as they break down organic material. We were attempting to establish the nature of the carbon compounds in which *R. rubrum* first stores the energy of light. Reasoning from analogy with what was known about processes such as muscle contraction, we deduced that certain organic phosphate compounds might be involved. In the effort to identify and locate these compounds we supplied isotopically labeled (tracer) phosphate to actively metabolizing cells of *R. rubrum* under illumination.

We found, however, that when the organisms were grown

in the usual media, they produced large excess amounts of unstable phosphate which broke down during separation for chemical analysis and máde it impossible to tell where the labeled phosphate had originally been taken up. We therefore had the problem of finding a diet which would allow the organisms to make only the minimum amount of phosphate required for growth. Fortunately S. H. Hutner of the Haskins Labouratories in New York had developed a completely synthetic medium for the growth of *R. rubrum* which could be adapted to our purpose simply by reduction of its phosphate content.

The dread St. Louis summer was approaching, and to make the new experiment Gest repaired to the Hopkins Marine Station at Pacific Grove, California, where facilities had been placed at our disposal. He grew *R. rubrum* in the modified Hutner medium in glass-stoppered bottles. The source of carbon in the medium was the sodium salt of malic acid. When *R. rubrum* breaks down this malate to take carbon, it liberates the alkaline sodium and some carbon dioxide. Carbon dioxide is soluble in alkali; hence it should promptly be dissolved in the liquid.

But Gest found that considerable amounts of gas were forming in the stoppered bottles a thick froth appeared on the liquid. He quickly ascertained that the gas was hydrogen. The organisms were producing about as much hydrogen as carbon dioxide. This was altogether mystifying, for not only had the formation of hydrogen never been observed before during photosynthetic metabolism, but it was known that these bacteria possessed an active system for taking up molecular hydrogen in the presence of carbon dioxide. Although there were large amounts of carbon dioxide in the bottles, free molecular hydrogen was still being evolved.

In the autumn Gest and I tackled this mystery in our home labouratory at Washington University in St. Louis. What had begun as a routine project in phosphate nutrition had evolved quite unexpectedly into a totally different sort of research. We dropped the original inquiry in order to investigate the new phenomenon.

We soon discovered that when ammonia salts were used as a source of nitrogen for the bacteria, instead of the glutamic acid of Hutner's formula, hydrogen formation was inhibited. These observations made it easy to understand why hydrogen production had not been noted before in the many researches with cultures of *R. rubrum*. The nitrogen source customarily employed had been ammonia! Our inadvertent substitution of glutamic acid for ammonia, by adoption of Hutner's medium, had been responsible for the discovery of the ability of *R. rubrum* to produce hydrogen photochemically.

We learned that hydrogen production could be supported not only by glutamic acid but by other amino acids and a large variety of carbon compounds. As a test compound for further researches we fixed on malic acid, the acid in apples. We proceeded to try to measure the production of hydrogen and carbon by the bacteria, and for these measurements we used a manometer (a device for recording pressure).

The manometer vessel was filled, as is customary, with the inert gas nitrogen. We now experienced our second shock. Organisms which had been producing hydrogen quite vigourously in the usual bottle cultures stopped doing so on being transferred to the manometer vessels. At first we supposed that some contaminant in the nitrogen, probably traces of oxygen or carbon monoxide, was responsible. But after weeks of futile experimentation, involving labourious purification of the nitrogen, we concluded that no contaminating gas could account for the situation.

We finally decided to substitute other inert gases–helium and argon–for nitrogen in the manometer vessels. Now *R. rubrum* resumed producing hydrogen as before! We found further that the addition of a little nitrogen to the argon or helium promptly stopped their hydrogen production. In effect, then, molecular nitrogen appeared to be acting like ammonia with respect to the formation of hydrogen.

The fact that molecular nitrogen exerted such a profound effect on the metabolism of *R. rubrum* indicated strongly that this organism might fix nitrogen. Further experiments showed it not only might but did! This demonstration immediately generated the suspicion that many other photosynthetic bacteria might possess the same ability. It was known, for instance, that a purple sulfur bacterium belonging to the genus *Chromatium* produced hydrogen in a manner similar to that noted in *R. rubrum*, and hydrogen production appeared to be common to all photosynthetic bacteria.

We may pause in our chronicle at this point to remark on a fascinating historical sidelight provided in a recent article by H. Derx of the Treub Labouratory in Buitenzorg, Indonesia. Derx points out that when the Dutch bacteriologist M. W. Beijerinck discovered *Azotobacter* in 1901 he was struck by the remarkable similarities between his new organism and the photosynthetic bacterium Chromatium. Although Chromatium was not known to fix nitrogen, Beijerinck was strongly convinced that some intimate relationship between Chromatium and his new organisms would be found. We see now that his intuition was solidly based–both Azotobacter and Chromatium are nitrogen fixers. This is a good example of how the intuitions of great researchers such as Beijerinck probe deeper than they originally appear to, and of how discoveries science rarely fail to cast a shadow before them.

The sudden emergence of nitrogen fixation in a group of organisms so widely diversified and widely distributed as photosynthetic bacteria stimulated a re-examination of the classification of baceria. Many of the organisms classified as Azotobacter because of their nitrogen-fixing ability differ from typical Azotobacter species in appearance, in their pH requirements, in the types of cultures they form, and so on. Derx, noting the apparently widespread occurrence of the nitrogen-fixing property, argued that nitrogen fixation should, not be considered an overriding criterion for bacterial classification. He proposed that on the basis of morphology and physiology some of these organisms should be classified as a new genus, for which he suggested the name *Beijerinckia*.

The number of bacteria found to be capable of nitrogen fixation steadily increases. At Wisconsin Wilson and E. D. Rosenblum studied Clostridium with sensitive tracer techniques and learned that the old contention that only a few species of this organism could fix nitrogen was untrue; of 15 species examined all but three actually do fix nitrogen. In 1951 F. D. Sisler and Claude E. ZoBell at the Scripps Institute of Oceanography in California obtained evidence of nitrogen fixation by the so-called "sulfate reducers"–organisms belonging to the genus Desulfovibrio. These bacteria can thrive only when oxygen is excluded from their environment.

They were first isolated by Beijerinck in 1895 from canal mud and soil, and have since been found in marine sediments, in the brine of oil wells at great depths, in the water at the bottom of gasoline storage tanks and in other widely assorted places. They are often blamed for the corrosion of iron conduits, the breakdown of concrete, the destruction of large numbers of fish in the ocean and other damaging activities. These organisms make sulfide, sulfur and organic sulfur compounds as products of their metabolism.

The cultures have the odor of bad eggs. Investigating the utilization of hydrogen by these organisms, Sisler and ZoBell incubated the organisms in glass-stoppered bottles of sea water with hydrogen gas in the space over the liquid. They also used a control flask with nitrogen instead of hydrogen in this space. The nitrogen disappeared from the gas phase at a rate which could not be accounted for by solution or diffusion. Growth experiments with molecular nitrogen confirmed that nitrogen fixation was taking place. Thus Desulfovibrio turned out to be not such a bad egg after all! Probably many more nitrogen fixers remain to be discovered. The number already found presents us with a radically altered picture of potentialities in the operation of the nitrogen cycle. The rootinhabiting Rhizobia and other land organisms with which we have long been acquainted undoubtedly play a major role in maintaining the cycle, but it now appears that the oceans also may be vast reservoirs for nitrogen fixation.

Possibly the organisms in the seas, including blue-green algae, fix even more nitrogen than do those in the world's soil. A great deal of nitrogen fixation may also be going on in tropical jungles, swamps and lakes. It has long been known that rice fields, for example, may remain fertile for long periods without fertilization. The Indian investigator P. K. De showed in 1988 that certain blue-green algae living in the soil of rice fields in India were very active nitrogen fixers, and concluded that they were probably the main agents in maintaining the fertility of the fields.

This classic demonstration of the importance of the blue-green algae has since been reinforced by work on the physiology and nutritional characteristics of a wide variety of them. These algae, by virtue of being both photosynthetic and nitrogen-fixing organisms, can thrive under conditions that permit no other form of life. That this is so has been shown in a number of instances. They are often the first organisms to colonize areas denuded of life, such as bare rock and soil. Within a few years after the volcanic explosion of Krakatoa in 1883, for instance, the pumice and volcanic ash had been repopulated by these dark green, gelatinous algae. Blue-green algae also are believed to contribute importantly to the production of living materials in fresh water–a fact which harasses resort operators at scenic lakes, which are often covered with profuse growths of noxious "blooms" as a result of the activities of these algae.

In hardiness and physiological versatility the photosynthetic bacteria are comparable to the algae. It may be, therefore, that the fertility of the Indian rice fields is not due solely to the algae, as De suggested, but also in part to photosynthetic bacteria. Other potential sites of nitrogen fixation are the environments in which the sulfate-reducing organisms like to live. This is not to say that every concrete piling or iron conduit sunk in the earth becomes an inadvertent possible source of agricultural fertilization, but man may very well learn how to take useful advantage of what he learns about the capacities of these organisms.

The discovery that nitrogen can be fixed by so many more organisms than we had suspected opens up exciting vistas. We can look forward to the possibility that we may some day be able to exploit the power of these organisms, just as we have already done with Rhizobia, and so help nature's nitrogen cycle to enrich our earth.

Chapter 3

Properties of Soils

SOIL PHYSICAL PROPERTIES

MINERAL PARTICLES

The principal properties of soil mineral particles in an environmental context are their size, shape, nature of surface, orientation and mineralogy. The mineral fraction of soils consists of particles that vary dramatically in size from large boulders, through cobbles and pebbles (several cm in diameter) to sand, silt and clay (<2 mm in diameter). Most studies of soil are concerned with the <2 mm size range, which is often referred to as the *fine fraction* or *fine earth.* It is the proportion by weight of the size categories within the fine fraction which defines the particle size distribution or *texture* of a soil. Examples of commonly used textural classification systems.

The particle size distribution of a soil is usually recorded numerically on a percentage by weight basis, or graphically using either a triangular graph, on which only three size categories can be shown, or a cumulative frequency curve which allows a fuller range of size classes to be recorded. Most soils comprise a continuous spectrum of particle sizes, and the width of this spectrum is defined by the degree of *sorting*. Poorly sorted soils possess a wide range of particle sizes, whereas well sorted soils have a narrow range.

Texture can be estimated in the field simply by rubbing the soil between thumb and forefinger. Sand grains are easily distinguished by their coarseness, while silt has a distinctive soapy feel and clay is characteristically plastic and mouldable when moist. For a more detailed indication of texture, however, labouratory analysis of the soil is required. This can be effected by the traditional method which involves sieving and sedimentation, or by more rapid methods using specialised instrumentation such as a Coulter Counter or laser diffractometer.

The morphology of mineral particles relates to their shape and surface characteristics. For large particles this can be determined in the field, but for smaller material it is necessary to use a microscope. Particle shape can be expressed in two or three dimensions. Common terms adopted in two-dimensional expression include roundness, angularity and elongation, while in three-dimensional expression particles are recorded in terms of the extent of sphericity and flatness, using descriptions such as sphere, disc, rod, cube and prism.

Such descriptions can be qualitative, or by reference to a comparison chart semi-quantitative estimates can be made. A number of indices have also been devised in order to quantify shape more accurately. The nature of the surface of a particle is usually referred to as its *surface texture,* although this in no way relates to the texture of a soil as an expression of its particle size distribution. Surface texture is most commonly evaluated for the fine fraction, using a scanning electron microscope, which allows the grain surfaces to be viewed in three dimensions at very high magnification. Specific surface features, or combinations of features, are often characteristic of a particular genetic environment.

Orientation of mineral particles describes the disposition of the particle in three dimensions, and this can be a useful indicator of the direction of movement of soil material. The measurement of orientation is usually made with reference to particle longest axes, either in the horizontal plane as a compass bearing, or in the vertical plane as an angle of dip or plunge. In some cases, both sets of measurement are made and the results combined. Orientation can be depicted in a variety of ways including bar charts, rose diagrams and polar scattergrams.

Minerals differ markedly in their composition and consequently many physical and chemical characteristics of soils, including texture, acidity and nutrient status, can be related to their mineralogy. Mineralogical studies are also important in the examination of weathering in soils. The mineralogical characteristics of large particles can be evaluated in the field by the inspection of hand specimens, but for smaller particles labouratory methods are required. Sand- and coarse silt-sized particles are usually examined using a petrological microscope, but clay and fine silt are too small to be easily resolved in this way and their mineralogy is commonly identified using the technique of X-ray diffractometry. This allows the crystal lattice dimensions to be measured and is particularly useful in the identification of sheet silicate minerals.

Samples are scanned by an X-ray beam and when a dominant lattice spacing is encountered, diffraction occurs and a peak is registered on a moving chart recorder. Each mineral has a characteristic set of peaks, although in some cases more than one mineral may have a peak in the same position. However, it is possible to distinguish between these minerals using simple chemical or heat pre-treatments which produce characteristic changes in the lattice spacing of one mineral but not in another.

AGGREGATES

Aggregation in soils is promoted by a number of physical, chemical and biotic forces. Physical forces include expansion and shrinkage associated with wetting and drying, and compaction by raindrop impact, animal trampling and agricultural machinery.

(a) cation bridging, (b) edge-face attraction of kaolinite crystals, (c) attraction between kaolinite crystals and an organic polyanion to form a 'string of beads' arrangement Chemical forces are largely electrostatic in character and often depend on the presence of adsorbed cations in association with the negative surface charge of colloidal particles such as clay and humus.

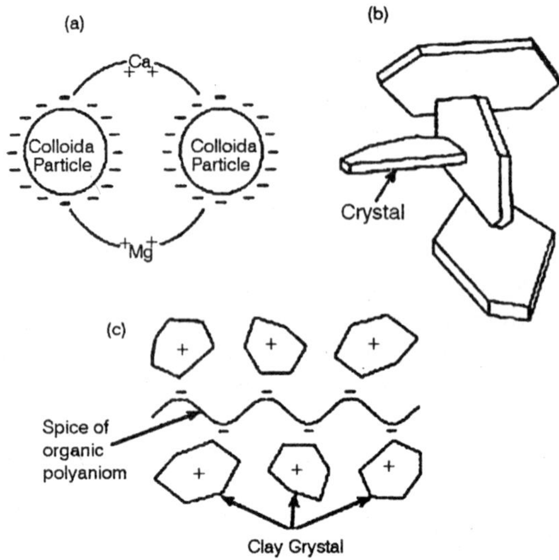

Fig. Interparticle Attraction in Aggregate Formation

Multivalent cations in particular, such as Ca^{2+}, Mg^{2+} and Al^{3+}, have the ability to form an attachment with more than one colloidal particle, a process known as *cation bridging*. Similarly, attraction may occur between positive charges on the broken edges of sheet silicate particles and the negative charges on the faces of other similar particles. In soils with significant positive charge, anion bridging may occur, particularly if multivalent anions are present.

In response to the various mechanisms of interparticle attraction, particles flocculate together to form larger units known as *domains;* these may be up to 5 ìm in diameter. In addition to electrostatic interparticle forces, interaggregate attraction can occur due to the binding effects of various organic compounds, fungal hyphae and plant roots.

Organic polymers, such as polysaccharide gums, together with organic mucilages and fungal hyphae, facilitate the binding of domains to form microaggregates which may be up to 250 ìm in diameter. At the larger scale, plant roots and fungal hyphae play an important role in the binding of microaggregates to form macroaggregates or *peds* which can be easily observed in the field. Inorganic cementing agents such as carbonate and iron compounds can perform a similar binding function in soils.

Because of the different scales of aggregation, it is often viewed in terms of a hierarchical model comprising building blocks which increase progressively in size. The degree of persistence of aggregates varies between the different levels in the model. Electrostatic forces predominate at the domain level and are particularly resistant to change. Similarly, at the microaggregate level, binding forces are relatively persistent, although they vary to some extent according to the organic content of the soil. In contrast, at the macroaggregate level, binding forces are often transient, being closely influenced by variations in plant cover and the development of root networks.

Aggregates, or peds, which persist during wetting/drying and freezing/ thawing cycles form the basis of soil *structure*. Soil structure is defined in terms of the size, shape and arrangement of particles, aggregates and pores. It is classified on the basis of ped morphology (shape), class (size) and grade (distinctiveness and durability). There are four main groups of ped morphology—spheroidal, blocky, prismatic and platy. Spheroidal peds are equidimensional in form and can be divided into granular (non-porous) and crumb (porous) types. Granular and crumb structures are most commonly found in the A horizons of soils, their development being facilitated by the presence of roots and other forms of organic material. Blocky peds are also equidimensional and may possess several curved or planar surfaces.

Frequently they are observed in B horizons, particularly in soils with significant amounts of clay, where they develop in response to the effects of regular wetting and drying cycles. Prismatic and associated columnar peds are characterised by strong vertical and weak horizontal development. The tops of prismatic peds are poorly defined, whereas those of columnar peds are often rounded in form. Like blocky structures, prismatic and columnar structures often develop as a result of wetting and drying but are found in the B horizons of less clay-rich soils.

Platy structure displays strong horizontal and weak vertical development, and often develops as a result of compaction. The structures are sometimes lenticular in form, with their centres being thicker than their edges, and these are often found in soils that are subjected to regular freezing and thawing cycles. Peds which comprise more than one structural type are often observed in soils and are known as *compound peds*. For example, large prismatic peds may be made up of smaller, blocky peds. It is also not uncommon to find that structures vary between the horizons within a soil profile, such that the A, B and C horizons could possess crumb, prismatic and blocky structures respectively.

Ped class or size is usually classified according to the system. Grade, which describes the distinctiveness and durability of peds, may be expressed as structureless (apedal), or as weakly, moderately, well or strongly developed. An apedal soil is described as massive if it is coherent and as single grain if it is not. Usually, massive soils are fine-textured while single grain are coarse-textured. Weakly developed structure comprises poorly formed, indistinct and weakly coherent peds, most of which will be broken down if the soil is disturbed, while strongly developed structure comprises well formed, distinct and durable peds. The grade of structure depends on a number of soil characteristics, particularly moisture content and the quantity of aggregating colloids such as clay and humus; peds tend to be considerably more durable in dry soils and in those with high colloidal contents.

Pore Space

Pore spaces vary dramatically in shape from spherical voids to tortuous, interconnecting cracks and channels. They also vary in size from large macropores of several cm in diameter to very fine micropores which may be <1 μm in diameter. Generally, a diameter of 60 μm (0.06 mm) is taken as the size boundary between macro- and micropores. Pore space will influence both the *bulk density* and the *porosity* of a soil. Bulk density refers to the specific gravity of a bulk soil sample, usually collected as an undisturbed core.

Its calculation is then derived from measurement of the mass and volume of the dried core. Values of bulk density are generally considerably lower than those of the particles which make up a soil, because they are based on both solid material and pore space rather than on the solid components alone. For example, the average density of soil particles is often assumed to be 2.65 g cm^{-3}, whereas bulk density values can range from around 2.0 g cm^{-3} in sandy soils and compacted clay layers to <1.0 g cm^{-3} in organic soils. If the bulk density of a soil is not measured, then an average value of 1.33 g cm^{-3} is often assumed. Porosity is a measure of the percentage volume of pore space, and can be determined indirectly from particle and bulk density as follows

$$\text{Porosity}(\%)=1-\frac{\text{Bulk density}}{\text{Particle density}}\times 100$$

Consequently, a soil with a bulk density of 1.33 g cm^{-3} and a particle density of 2.65 g cm^{-3} will have a porosity of 50 percent. In some instances porosity is expressed as a ratio rather than a percentage, thus the pore space ratio (PSR) of a soil with 50 percent porosity is 0.5. Porosity can also be determined directly from an undisturbed core of known volume (V). Initially the core is saturated with water and the weight (Ws) recorded. It is then dried and re-weighed (Wd), and the difference between Ws and Wd represents the weight of water held by the core in its saturated state. This will be equivalent to the water volume, because the specific gravity of water is 1.0 g cm"3. Hence, porosity is calculated as follows

$$\text{Porosity}(\%)=\frac{\text{Ws}-Wd}{\text{V}}\times 100$$

Using undisturbed cores in their field moist state, it is possible to determine the proportions of water and air occupying the pore space. Water-filled porosity is equivalent to the volumetric water content (θv) of the soil. This is calculated from the weight of field moist soil (Wm), its weight after drying (Wd) and the volume of the core (V)

$$\theta\text{v}=\frac{\text{Ws}-Wd}{\text{V}}\times 100$$

As with total porosity, values of water-filled porosity can be expressed as a percentage or on a proportional basis. On average, values are about 25 percent or 0.25, although in a saturated soil the value will be equivalent to the total porosity, whereas in a dry soil it may be <5 percent or 0.05. Air-filled porosity can be calculated by subtraction of water-filled porosity from total porosity. A more sophisticated method of porosity measurement is provided by image analysis, in which undisturbed sections of soils are photographed and the total area occupied by pores is quantified using an image analyser.

In addition to porosity, pore size distribution is an important soil characteristic. It is associated directly with water retention, drainage and aeration, and therefore has a major influence on plant growth. It is most often established from moisture characteristic or moisture retention curves. This involves determining the volumetric water content at various points over a range of tensions or suctions applied to an undisturbed soil core. As tension is increased, water is removed from progressively smaller pores and because tension is inversely proportional to pore radius, the volume of pores of a certain size can be determined from the amount of water extracted at the appropriate tension. The total porosity is 50 percent and, of this, 15 percent of pores are > 15 μm in diameter, 15 percent are 1.5-15 μm, 10 percent are 0.15-1.5 μm and 10 percent are <0.15 μm.

Porosity and pore size distribution are influenced by a number of soil characteristics, particularly texture, degree of aggregation, bulk density, presence of swelling clays, and organic content. Closely packed sands, for example, could have a theoretical minimum porosity of about 10 percent. However, as soils usually possess a wide range of particle sizes and aggregates, and because soil particles are rarely spherical, values of porosity are usually much greater than this. Sandy and compacted clay soils may have porosities of <40 percent, whereas a fine-textured A horizon can have values of > 60 percent.

Moisture

Soil water possesses free energy which is a measure of its

potential for movement and change in the soil. In soils with a high moisture content, forces attracting the water to solid particles are weak and its free energy is high. As moisture content decreases, however, the attractive forces become progressively stronger and its free energy decreases. Soil moisture is affected by three types of force determined by soil properties which can either encourage or restrict water movement. First is *adsorption* whereby water molecules are attracted to the surfaces of colloids mainly by electrostatic forces.

This is therefore an important process in soils with high clay or organic matter contents, in which such forces are high. Second is *capillarity* by which water is held in soil pores by adsorptive forces between the water and pore surfaces and surface tension forces at the water surface. The force by which the water is held increases with decreasing pore size, therefore water can drain out of large pores more easily than from smaller ones. The combined effect of adsorptive forces and capillarity is known as *matric suction*. The third force is *osmosis* which occurs between solutions of different ionic concentrations, water moving from lower to higher concentration solutions. This is important in saline soils in which high solute concentrations can form due to the relative ease of dissolution of salts.

These forces are usually expressed in units of atmospheres (atm), bars or kiloPascals (kPa), with 1 atm being approximately equal to 1 bar (1.02 atm= 1 bar), and 1 bar equal to 100 kPa. In the past, these forces were sometimes expressed as the logarithm of the *hydraulic head* (pF); this is the length of a column of water which is required to produce a given positive pressure, or which can be supported by a given negative pressure (tension). This concept is, however, rarely used today.

Soil water is held most strongly (30 to > 1,000 atm) when it is adsorbed onto colloidal particle surfaces in the form of thin films only a few molecules in thickness (hygroscopic water), or when it is bound up in mineral structures (structural water). Further away from particle surfaces, water is held in small micropores at tensions of 0.05-30 atm by capillarity;

much of this water can be lost from the soil through evapouration and plant uptake. Water held at tensions below 0.05 atm occurs in the macropores, and this will usually drain rapidly from the soil, in two or three days, under the influence of gravity and is therefore known as *gravitational water.*

Thus, on moving away from particle surfaces into progressively larger pores, tensional forces become progressively weaker, decreasing logarithmically with increasing pore size, until they are unable to counteract the effect of gravity. A soil which has lost all of its gravitational water through drainage contains the maximum level of plant-available water and is said to be at *field capacity*. If, however, a soil loses all of its available water through evapouration and plant uptake, without further wetting, then it is said to have reached *wilting point*.

A variety of methods are available for the measurement of soil moisture. For example, it can be determined gravimetrically using bulk samples; these are weighed in their field-moist and oven-dried states, the weight difference representing the moisture content, which is then expressed as a percentage of either field-moist or oven-dried soil. Alternatively, moisture content can be expressed on a volumetric basis using soil cores of known volume. Weight measurements are made as above and, given that the specific gravity of water is 1.0 g cm^{-3}, the moisture content can then be expressed as a percentage of the soil volume.

Changes in moisture content within a soil can be determined using a neutron probe. As the probe is lowered through aluminium tubes inserted into the soil, it emits neutrons which collide with the hydrogen nuclei contained in water molecules, and a sensor detects backscatter from the collisions, the intensity of which is in direct proportion to water content. Problems have, however, been experienced with the use of neutron probes, particularly in soils containing significant amounts of swelling clays, where cracking results from drying and associated shrinkage.

The strength of tensional forces (matric potential) by which water is held in the soil can be determined using a tensiometer.

This consists of a sealed plastic tube filled with water, with a ceramic cup at one end and a vacuum gauge at the other. The tube is set in the soil at the required depth and the gauge adjusted to zero. As water moves out into the soil through the cup, tension increases progressively until the water in the tensiometer is in equilibrium with that in the soil. At this point, movement ceases and the tension recorded on the gauge is a measure of the matric potential; tensiometers can also be connected to pressure transducers and logged electronically. The tensiometer can only be used at relatively low tensions, between about 0 and "1.0 atm, but this range is important within the context of plant growth; plants obtain water via their roots by exerting a tensional force of about 0.05-15 atm.

Temperature

Soil temperature is an extremely dynamic property, varying diurnally and seasonally, with the effects being most rapid and extreme towards the surface. On a diurnal time-scale, soils are heated during the day and the effect gradually extends downwards, perhaps taking several hours to reach a depth of 30 cm. At night soils cool rapidly at the surface and heat is transferred upwards from within the soil. Seasonal heating and cooling cycles operate in a similar manner, but they penetrate deeper into the soil than diurnal cycles because the time-scale is much greater; diurnal cycles usually affect only the upper 30 cm or so, whereas seasonal cycles can penetrate to a depth of several metres.

In order to detect their rapid variation, soil temperatures must be measured frequently and at a variety of depths, and are usually recorded electronically using a series of thermistors linked to an automatic recorder. Soil temperature is influenced by a number of soil properties, in particular texture, moisture and organic content; these will be discussed later in the context of soil influences on small-scale climatic characteristics.

Mechanics

Commonly encountered soil mechanical properties are strength, stability and consistence. The mechanical strength and stability of soils are derived from interparticle and interped forces responsible for the development of soil structure. Aggregate stability is a useful measure of the structural stability of soils, and is based on the ability of aggregates to survive wetting. During wetting, air is trapped in pores within the peds, and the pressure of the trapped air increases until it is sufficient to cause aggregate breakdown or *slaking*. In addition, water entering the aggregates interferes with electrostatic forces of interparticle attraction and may dissolve cementing agents, thus causing further weakening.

Measurements of aggregate stability have included determining the proportion by weight of aggregates retained on a 2 mm sieve after wet-sieving or after ultrasonic dispersion, although such methods have been questioned because they subject the soil to artificial forces for relatively short periods of time. The strength and stability of a soil can also be assessed from its resistance to compression and shear, which provides a useful indication of the degree of cohesion. Resistance to compression can be determined using a penetrometer, whilst a shear vane can be used to measure resistance to shear. These characteristics can also be established from a triaxial shear test, in which a soil core is first subjected to compression in order to determine its load-bearing capacity, and then to a shear test.

Soil strength is closely related to a number of soil properties, particularly texture, organic content, bulk density and moisture content. Coarse-textured soils frequently possess a relatively high degree of strength due to the often irregular nature of particle boundaries and the resulting large surface area of contact. Coarse-textured soils with smooth particle boundaries are considerably weaker, however, due to the smaller surface area of contact. Fine-textured soils are often strongly cohesive and well aggregated, and are particularly strong when dry. When wet, however, the cohesive forces tend to break down and soil strength decreases dramatically. The presence of organic matter improves the cohesive properties of soils and increases their strength. Similarly, soil strength increases with increasing bulk density, due argely to the reduction in total porosity.

In contrast to cohesion, dispersion may reduce the strength and stability of soils. This commonly occurs in soils whose colloidal fraction is dominated by adsorbed sodium, as often occurs in semi-arid environments. Sodium ions are only loosely held by colloidal particles, and are unable to neutralise fully the surface negative charge. This situation promotes interparticle repulsion rather than attraction, thus resulting in breakdown of soil structure. Structure may also be disrupted by repeated shrinkage and swelling, which can be particularly active in soils containing large amounts of swelling clays. The shrink-swell potential of a soil can be determined from the coefficient of linear extensibility (COLE), which describes the volume change of a length of soil on drying, and is calculated as follows

$$\text{COLE} = \frac{\text{Lm-Ld}}{\text{Ld}}$$

Where Lm and Ld are the moist and dry sample lengths respectively.

Soil consistence describes the change in state of soil, from solid through plastic to liquid, with increasing moisture content. It is established from the three Atterberg limits—shrinkage, plastic and liquid limits. The shrinkage limit represents the minimum moisture content above which soil volume begins to increase with further wetting; below this limit soil volume remains constant irrespective of changes in moisture content. The plastic limit represents the minimum moisture content above which the soil becomes mouldable, and the liquid limit represents the minimum moisture content above which the soil flows under its own weight. The plastic and liquid limits in particular are closely associated with a number of soil properties, for example cation exchange capacity, clay content and specific surface area.

Colour

Soil colour is a characteristic which can easily be determined in the field, and which provides useful information regarding the presence or absence of certain soil constituents. For example, dark colours are usually indicative of high organic contents, while red colours are characteristic of soil rich in iron oxides, and blue-grey colours indicate the presence of iron in its reduced form. Such indications are, however, only general because, for example, manganese can also produce a dark colouration, while the intensity of colour is often related to moisture content. The identification of soil colour is also rather subjective, as different individuals interpret colour in different ways.

In an attempt to reduce subjectivity, soil colour is most often determined using the Munsell system where soil samples are matched against standard colour charts. Munsell colour notation has three components—the *hue* which indicates the major colour(s) present, the *value* which is a measure of the degree of darkness or lightness of the colour, and the *chroma* which is a measure of colour intensity. A sample of soil is therefore represented by a notation according to the colour to which it most closely corresponds, and each notation has an equivalent description; for example, a soil with a hue of 10YR, a value of 3 and a chroma of 4 is represented as 10YR 3/4, which has the description yellowish brown. The objective determination of soil colour has been the subject of much discussion, and although the Munsell system is the most widely adopted, other methods and indices have been developed.

SOIL CHEMICAL PROPERTIES

ELEMENTS AND COMPOUNDS

Elements and compounds in a soil occur in two principal forms—as the chemicals that make up the structure of the basic soil constituents, and as individual components which are held in the soil by interparticle attraction. The first of these are important in terms of soil properties only once the constituents start to break down, whereupon they are released into the soil, but the second of these are generally of more immediate importance because they are more readily available for interaction. The chemicals that make up the structure of mineral material are determined by total chemical analysis, for example by atomic absorption spectrometry following dissolution of the material in strong acids, or by the more rapid method of X-ray fluorescence spectrometry.

In terms of organic matter, the principal structural elements which are normally determined are carbon and nitrogen. Organic carbon content is usually determined by dichromate oxidation, and this can also be used to estimate the organic matter content of a soil by using a conversion factor of 1.7 (organic matter %=organic carbon %×1.7). Organic matter content can also be determined by loss on ignition or by digestion with hydrogen peroxide, although these methods may not always be particularly accurate. Nitrogen content is most often determined by the Kjeldahl digestion procedure.

Individual elements and compounds which are commonly examined as individual components include exchangeable bases, free iron and aluminium, carbonates and heavy metals. Exchangeable bases are held in the ion exchange complex of a soil and are usually extracted using ammonium acetate, then analysed by flame emission or atomic absorption spectrometric methods. Exchangeable cation content is commonly expressed in units of milliequivalents (me) per 100 g of dry soil, which represents the amount of a cation that will replace or combine with 1 mg of hydrogen per 100 g of dry soil.

Exchangeable base content can also be expressed in units of cmol(+)kg^{-1}, which is exactly the same as me per 100 g. Free iron and aluminium are present in soils in a number of forms, including sesquioxides (oxide and hydroxide compounds), poorly crystalline or amorphous allophanic and imogolitic materials, and organometallic complexes. These can be determined using a range of selective extractants. For example, total free iron and aluminium is often determined by extraction with sodium dithionite or by a dithionite-citrate-bicarbonate (DCB) extraction. Inorganic forms of iron and aluminium, which are poorly crystalline or amorphous, can be extracted by sodium oxalate, and organically bound forms by potassium pyrophosphate.

The potency of these extractions decreases in the order DCB> oxalate> pyrophosphate, and there is thought to be relatively little overlap between them. It is therefore possible to determine levels of all these different forms by sequential extraction performed on the same sample. The most common carbonates in soils are compounds of calcium and magnesium, derived largely from carbonate-rich parent materials. Presence or absence of carbonates can be established simply by adding a few drops of dilute hydrochloric acid (HCl) to a small sample of soil; effervescence is observed if carbonates are present. More accurate measurement can be made in the labouratory, for example using a calcimeter, which measures the calcium carbonate ($CaCO_3$) equivalent by determining the quantity of carbon dioxide (CO_2) evolved during treatment with hydrochloric acid

$$2HCl + CaCO_3 \Rightarrow CaCl_2 + CO_2 + H_2O$$

Heavy metals can occur naturally in soils, but often occur at enhanced levels due to additions from motor vehicle emissions, sewage sludge applications, metal mining and smelting, and scrap metal processing. They can exist in a number of forms including numerous compounds, particularly oxides, sulphides and sulphates, metal cations such as lead, zinc and cadmium (Pb^{2+}, Zn^{2+} and Cd^{2+}), which may be adsorbed onto the surfaces of negatively charged colloidal particles, and organo-metallic complexes. Heavy metals are commonly extracted from soils by acid digestion, using a strong mineral acid such as nitric acid to determine total contents, and EDTA or a weak organic acid such as acetic acid to determine 'plant-available' levels. Contents are then measured by atomic absorption spectrometry.

Ion Exchange

Ion exchange is a most important soil property in that it plays a key role in plant nutrition, and in a broader context, in the development of many chemical characteristics of soils. Central to ion exchange is the way in which ions are held on the surfaces of colloidal particles. Colloidal material consists mainly of clay and humus particles and is often referred to as the *clay-humus complex* or *exchange complex*. These particles have a high specific surface area (ratio of surface area to volume) and possess both high surface energy and significant surface charge. This charge is largely negative and can occur either as permanent charge or variable (pH dependent) charge.

In the case of permanent charge, cations in mineral structures are replaced by cations of similar size but with lower valency (isomorphous substitution); such charge is independent of acidity. In the case of variable charge, hydrogen ions undergo reversible dissociation from surface groups such as –OH and $2H^+$ on the edges of clay minerals and oxides, and –COOH, –OH and $-NH_2$ in organic material.

As a result of their negative surface charge, colloidal particles behave like giant anions and are known as *micelles*. The micelles are able to attract cations onto their surfaces, a process known as *adsorption*. With increasing distance from the micelle surface, the concentration of cations decreases exponentially whilst the concentration of anions shows a reciprocal increase. The zone of adsorbed cations, together with the surface negative charge on the micelle, are often referred to as the *electrical double layer*.

The ease with which a cation is adsorbed depends on its valency and degree of hydration.

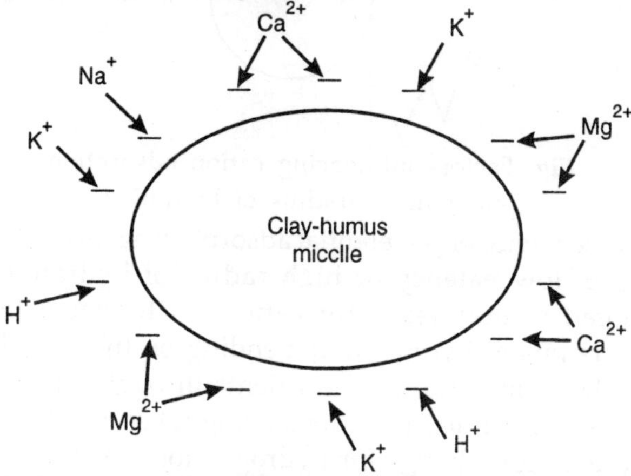

Fig Clay-humus Micelles and cation Adsorption

Cations with a high valency have a high energy of adsorption and are therefore adsorbed in preference to lower valency cations. For cations with equal valency, however, the one with the smallest *radius of hydration* (radius of the ion together with the associated water molecules surrounding it) is adsorbed preferentially because it can get closer to the micelle surface. The generally accepted sequence of preferential adsorption for base cations is $Ca^{2+} > Mg^{2+} > K^{+} > Na^{+}$, and this is reflected in their usual proportions in the soil (Ca^{2+}=80%, Mg^{2+}=15%, K^{+}+Na^{+} =5%). Adsorbed cations can be exchanged for cations in the soil solution

$$Micelle - A + B^{+}(aq) \Rightarrow Micelle - B + A^{+}(aq)$$

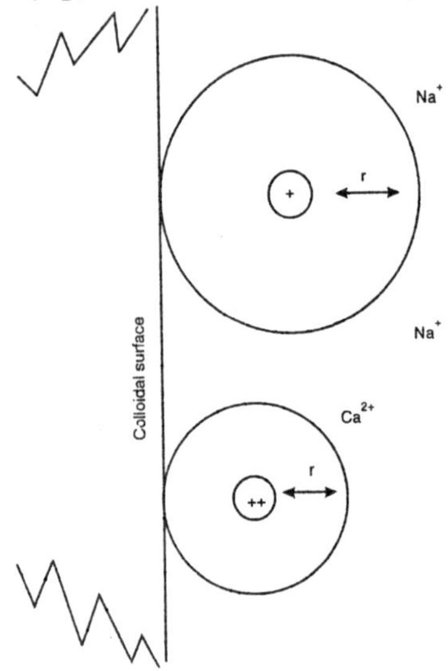

Fig. Factors influencing cation adsorption preference; r=radius of hydration

The sequence of preferential adsorption again applies, with cations of low valency or high radius of hydration being exchanged in preference for cations of higher valency or smaller radius of hydration, depending on their availability. Plants obtain many of their nutrients through this exchange mechanism, the nutrient cations being taken up through the root system in exchange for hydrogen ions. Although cation adsorption and exchange tend to predominate in soils, in some circumstances anion adsorption and exchange are more common. While the former occur in soil containing significant amounts of humus, and clay minerals with a high specific surface area such as montmorillonite, vermiculite and illite, the latter occur in soil with limited humus content and with clay minerals with a low specific surface area such as kaolinite. Oxides of iron and aluminium are also significant in anion adsorption as they possess variable surface charge which is positive under acidic conditions. This positive charge results from the combination of hydrogen (H^+) ions with edge hydroxyl groups, for example

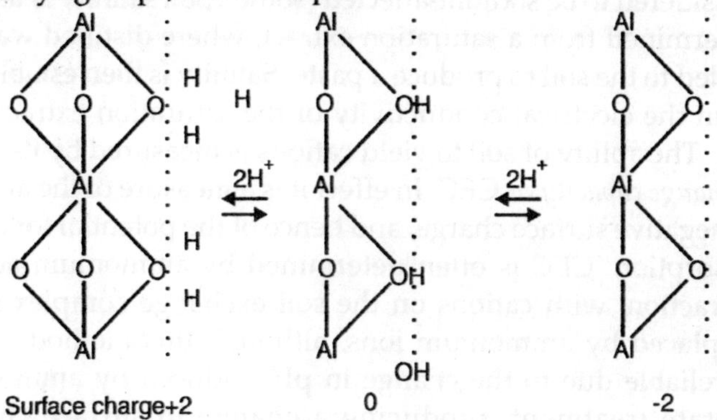

Fig. Influence of pH on Surface Charge in an Aluminium Oxide.

The addition of H^+ under acid Conditions Produces Positive Charge, whereas the Removal of H^+ at Higher pH Produces Negative Charge

$$Al\text{-}OH + H^+ \Rightarrow Al\text{-}OH_2^+$$

As in the case of cations, anions with a high valency are able to get closer to surfaces than anions with a low valency. Thus sulphate (SO_4^{2-}) and phosphate (PO_4^{3-}) are often strongly held, whereas nitrate (NO_3^-) tends to be held rather weakly. Like cations, adsorbed anions can also be exchanged for anions in the soil solution. Unlike Ca^{2+}, Mg^{2+} and K^+, which are important plant nutrients, Na^+ is toxic to many plant species, and it also has a deleterious effect on soil structure, promoting the dispersal of aggregates. The assessment of soil salinity can therefore be important in these instances, and various expressions are commonly used. The *sodium adsorption ratio* (SAR) represents the amount of Na^+ relative to Ca^{2+} and Mg^{2+}, and is determined as follows

$$SAR=\frac{N_a^+}{\left(Ca^{2+}+Mg^{2+}\right)^{0.5}}$$

The SAR is approximately equal to the *exchangeable sodium percentage* (ESP) of the soil, where the Na^+ level, measured in me per 100 g, is expressed as a percentage of the total exchangeable base content. An ESP value of 15 percent is generally accepted as the threshold above which soils are considered to be sodium-affected (sodic). Soil salinity is usually determined from a saturation extract, where distilled water is added to the soil to produce a paste. Salinity is then established from the electrical conductivity of the saturation extract.

The ability of soil to yield cations is measured by its *cation exchange capacity* or CEC. In effect it is a measure of the amount of negative surface charge, and hence of the potential for cation adsorption. CEC is often determined by ammonium acetate extraction, with cations on the soil exchange complex being displaced by ammonium ions, although this method can be unreliable due to the change in pH, induced by ammonium acetate treatment, producing a change in surface charge; alternative extraction methods have therefore been developed, but these have not yet become widely used.

As with exchangeable base content, CEC is expressed in units of me per 100 g or cmol(+)$kg^{''1}$. Values vary dramatically and tend to be highest in soils with high clay and organic contents. CEC values for organic matter may be 150-300 me per 100 g, virtually twice that of clay. Clay mineralogy also has a major influence on CEC, in terms of the charge density per unit area; CEC values range from 5-10 me per 100 g for kaolinite, through 30-40 for illite, to > 100 for vermiculite and montmorillonite.

From the content of exchangeable bases and CEC, it is possible to determine the proportion of exchangeable bases on the soil exchange complex, a property known as the *percentage base saturation*. This is calculated as follows

$$\text{Base saturation}(\%)=\frac{\left(Ca^{2+}+Mg^{2+}+K^{+}+Na^{+}\right)}{CEC}\times 100$$

The difference between CEC and total exchangeable base content provides an approximate measure of exchangeable hydrogen content, or *exchangeable acidity*

$$\text{Exchangeable}\,H^{+} = \text{CEC-}\left(Ca^{2+} + Mg^{2+} + K^{+} + Na^{+}\right)$$

It is also possible to measure the *anion exchange capacity* (AEC) of a soil. Reliable methods of determination have been developed in relation to tropical soils, which often possess significant AEC values, unlike soils of temperate environments. Essentially these methods are based on the degree of adsorption of an index anion, usually Cl^{-} or NO_3.

Acidity and pH

Acids in aqueous solutions undergo dissociation to release their constituent ions, namely hydrogen (H^{+}) and an acid anion, for example

$$\underset{\text{(sulphuric acid)}}{H_2SO_4} \Rightarrow 2H^{+} + \underset{\text{(sulphate)}}{SO_4^{2-}}$$

Acidity is measured in terms of the H^{+} ion concentration using the pH scale. The relationship between pH and H^{+} ion concentration is inverse and logarithmic

$$p^{H} = -\log_{10}\left[H^{+}\right]$$

The pH scale ranges from 1.0 at the most acidic extreme to 14.0 at the alkaline extreme, with a value of 7.0 at neutrality. The pH of soils varies widely, from around 2.0 in acid sulphate soils to about 12.0 in alkaline sodic soils; good quality agricultural soils have a value around 6.0 to 7.0.

The measurement of soil pH is usually made in a standard suspension of 1 2.5 weight to volume (e.g. 10 g of soil in 25 ml distilled water), in order to ensure data comparability. Although distilled water is often used to make up the suspension, a suspension made with a dilute solution of calcium chloride (0.01M) is sometimes used in order to provide a more realistic value of H^{+} concentration by minimising calcium release from the soil exchange complex. For this reason pH levels measured in calcium chloride suspension are generally lower than those recorded in a suspension made up with distilled water. The measurement itself can be made using electrometric (pH probe) or colourimetric techniques.

In addition to H^+ ions, Al^{3+} ions play an important role in the generation of soil acidity, particularly in soils that are already acidic. Al^{3+} ions undergo hydrolysis during which H^+ ions are released into the soil solution

$$Al^3 + H_2O \Rightarrow Al(OH)_2^+ + H^+$$

$$(\text{hydroxy-aluminium})$$

Positively charged hydroxy-aluminium species can then occupy exchange sites, thus resulting in reduced CEC. Hydroxy-aluminium species may undergo further hydrolysis to produce yet more H^+ ions and stable aluminium hydroxide (gibbsite)

$$Al(OH)_2^+ + 2H_20 \Rightarrow Al(OH)_3 + H^+$$

(hydroxy-aluminium)

Consequently, soil acidity promotes the development of further acidity through aluminium hydrolysis, and this becomes an important source of H^+ ions when soils become acidic. If soil pH falls below about 5.5, Al^{3+} ions themselves begin to occupy exchange sites. Because of their higher valency, Al^{3+} ions are adsorbed much more strongly than divalent and monovalent cations, therefore levels of exchangeable aluminium increase, and amounts of exchangeable bases decrease, as pH declines. Soil acidity is closely related to many other soil properties such as organic content, exchangeable base content and CEC.

Aeration

Soil aeration relates to the amount of oxygen present in the soil atmosphere. This can be assessed using both direct and indirect methods. The rate of oxygen diffusion through the soil can be measured directly using a platinum electrode, while the concentration of oxygen in a sample of soil air can be determined by gas-liquid chromatography. A number of instruments are available for sampling the soil atmosphere and for monitoring changes in aeration, although problems have been experienced in their design, due largely to leakage.

Indirect assessment of soil aeration can be made from detection of fermentation and putrefaction products, either by quantitative analysis or simply by their odour. Limited aeration can also be seen by the appearance of reduced compounds of manganese and iron, which are characterised by the presence of black specks and pallid grey colours respectively. A particularly useful indicator of the degree of soil aeration is the *redox potential* (Eh) or oxidation-reduction status. Redox or oxidation-reduction reactions are those in which a chemical species undergoes oxidation or reduction through the transfer of electrons (e^-). An example of such a reaction is the reduction of ferric (Fe III) hydroxide to ferrous (Fe II) hydroxide

$$Fe(OH)_3\ e^- \Leftrightarrow Fe(OH)_2 + OH^-$$

(ferric hydroxide) (ferrous hydroxide)

This reaction is reversible, with a tendency to change towards the left under oxidising conditions and towards the right under reducing conditions. The redox potential (Eh) is the difference in electrical potential between the two halves of this coupled reaction. It can be measured using a platinum electrode and an appropriate reference electrode; these are inserted into the soil and the electrical potential difference (V) is recorded, and referred to a standard pH value of 7.0. Eh values in aerobic soils are generally between 0.3 and 0.8 V, while in anaerobic soils they are often between 0.3 and $^-$0.4 V.

Under aerobic conditions, electrons produced during respiration combine with oxygen. Under anaerobic conditions, however, oxygen is unavailable and other chemical species must act as electron receptors. Ferric iron compounds often take on this role, undergoing reduction to ferrous iron compounds. Each chemical species has a threshold redox potential below which it begins to act as an electron receptor and thus becomes unstable. Ferrous iron compounds are often present in the centre of peds where Eh may be 0.1 V lower than at the edges, whereas ferric compounds are more likely to occur between peds, in areas adjacent to root channels, and in patches where texture is relatively coarse.

Such microscale variation in distribution of the different

iron compounds can often give the soil a mottled appearance, where blue-grey colours of ferrous compounds are interspersed with orange-brown colours of ferric compounds. The degree of soil aeration is therefore closely associated with porosity and pore size distribution, and is inversely related to soil water content. It is also related to texture and the degree of structural development. Optimum levels of aeration are most often found in well structured soils with a fine loamy texture.

Chapter 4

Seed Chemistry

A seed is a small embryonic plant enclosed in a covering called the seed coat, usually with some stored food. It is the product of the ripened ovule of gymnosperm and angiosperm plants which occurs after fertilization and some growth within the mother plant. The formation of the seed completes the process of reproduction in seed plants (started with the development of flowers and pollination), with the embryo developed from the zygote and the seed coat from the integuments of the ovule.

Seeds have been an important development in the reproduction and spread of flowering plants, relative to more primitive plants like mosses, ferns and liverworts, which do not have seeds and use other means to propagate themselves. This can be seen by the success of seed plants (both gymnosperms and angiosperms) in dominating biological niches on land, from forests to grasslands both in hot and cold climates.

The term *seed* also has a general meaning that predates the above — anything that can be sown i.e. "seed" potatoes, "seeds" of corn or sunflower "seeds". In the case of sunflower and corn "seeds", what is sown is the seed enclosed in a shell or hull, and the potato is a tuber.

SEED STRUCTURE

A typical seed includes three basic parts

- An embryo,
- A supply of nutrients for the embryo,
- A seed coat.

The embryo is an immature plant from which a new plant will grow under proper conditions. The embryo has one cotyledon or seed leaf in monocotyledons, two cotyledons in almost all dicotyledons and two or more in gymnosperms. The radicle is the embryonic root. The plumule is the embryonic shoot. The embryonic stem above the point of attachment of the cotyledon(s) is the epicotyl. The embryonic stem below the point of attachment is the hypocotyl.

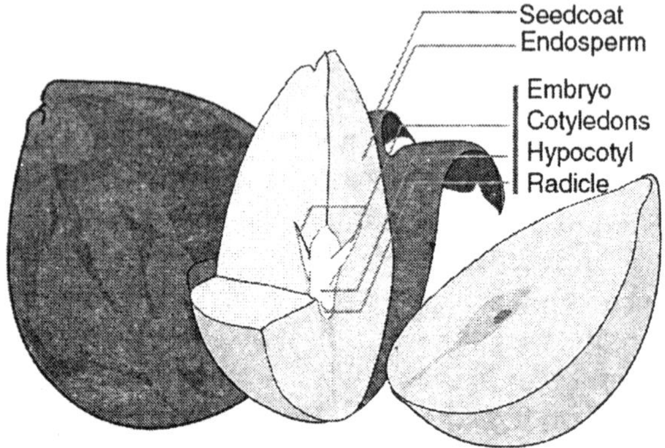

Fig. The Parts of an Avocado Seed (a dicot), Showing the seed Coat, endosperm, and embryo.

Within the seed, there usually is a store of nutrients for the seedling that will grow from the embryo. The form of the stored nutrition varies depending on the kind of plant. In angiosperms, the stored food begins as a tissue called the endosperm, which is derived from the parent plant via double fertilization. The usually triploid endosperm is rich in oil or starch and protein.

In gymnosperms, such as conifers, the food storage tissue is part of the female gametophyte, a haploid tissue. In some species, the embryo is embedded in the endosperm or female gametophyte, which the seedling will use upon germination. In others, the endosperm is absorbed by the embryo as the latter grows within the developing seed, and the cotyledons of the embryo become filled with this stored food.

At maturity, seeds of these species have no endosperm and are termed exalbuminous seeds. Some exalbuminous seeds are bean, pea, oak, walnut, squash, sunflower, and radish. Seeds with an endosperm at maturity are termed albuminous seeds.

Most monocots (e.g. grasses and palms) and many dicots (e.g. brazil nut and castor bean) have albuminous seeds. All gymnosperm seeds are albuminous.

The seed coat (or *testa*) develops from the tissue, the integument, originally surrounding the ovule. The seed coat in the mature seed can be a paper-thin layer (e.g. peanut) or something more substantial (e.g. thick and hard in honey locust and coconut). The seed coat helps protect the embryo from mechanical injury and from drying out.

In addition to the three basic seed parts, some seeds have an appendage on the seed coat such an aril (as in yew and nutmeg) or an elaiosome (as in Corydalis) or hairs (as in cotton). There may also be a scar on the seed coat, called the *hilum;* it is where the seed was attached to the ovary wall by the funiculus.

SEED PRODUCTION

Seeds are produced in several related groups of plants, and their manner of production distinguishes the angiosperms ("enclosed seeds") from the gymnosperms ("naked seeds"). Angiosperm seeds are produced in a hard or fleshy (or with layers of both) structure called a fruit that encloses the seeds, hence the name.

In gymnosperms, no special structure develops to enclose the seeds, which begin their development "naked" on the bracts of cones. However, the seeds do become covered by the cone scales as they develop in some species of conifer.

Kinds of Seeds

There are a number of modifications to seeds by different groups of plants. One example is that of the so-called *stone* fruits (such as the peach), where a hardened fruit layer (the endocarp) surrounds the actual seed and is fused to it.

Many structures commonly referred to as "seeds" are actually dry fruits. Sunflower seeds are sold commercially while still enclosed within the hard wall of the fruit, which must be split open to reach the seed.

Seed Development

The inside of a *Ginkgo* seed, showing a well-developed embryo, nutritive tissue (megagametophyte), and a bit of the surrounding seed coat.

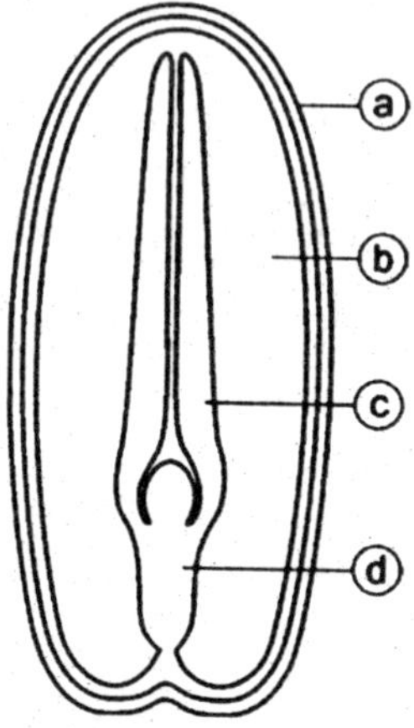

Fig. Diagram of the Internal Structure of a Dicot seed and Embryo. (a) Seed coat, (b) Endosperm, (c) Cotyledon, (d) Hypocotyl.

The seed, which is an embryo with two points of growth (one of which forms the stems the other the roots) is enclosed in a seed coat with some food reserves. Angiosperm seeds consist of three genetically distinct constituents

- The embryo formed from the zygote,
- The endosperm, which is normally triploid,
- The seed coat from tissue derived from the maternal tissue of the ovule.

In angiosperms, the process of seed development begins with double fertilization and involves the fusion of the egg and sperm nuclei into a zygote.

The second part of this process is the fusion of the polar nuclei with a second sperm cell nucleus, thus forming a primary endosperm. Right after fertilization the zygote is mostly inactive but the primary endosperm divides rapidly to form the endosperm tissue. This tissue becomes the food that the young plant will consume until the roots have developed after germination or it develops into a hard seed coat.

The seed coat forms from the two integuments or outer layers of cells of the ovule, which derive from tissue from the mother plant, the inner integument forms the tegmen and the outer forms the testa. When the seed coat forms from only one layer it is also called the testa, though not all such testa are homologous from one species to the next.

In gymnosperms, the two sperm cells transferred from the pollen do not develop seed by double fertilization but instead only one sperm fertilizes the egg while the other is not used. The seed is composed of the embryo (the result of fertilization) and tissue from the mother plant, which also form a cone around the seed in coniferous plants like Pine and Spruce.

The ovules after fertilization develop into the seeds; the main parts of the ovule are the funicle; which attaches the ovule to the placenta, the nucellus; the main region of the ovule were the embryo sac develops, the micropyle; A small pore or opening in the ovule where the pollen tube usually enters during the process of fertilization, and the chalaza; the base of the ovule opposite the micropyle, where integument and nucellus are joined together.

The shape of the ovules as they develop often affects the finale shape of the seeds. Plants generally produce ovules of four shapes the most common shape is called anatropous, with a curved shape. Orthotropous ovules are straight with all the parts of the ovule lined up in a long row producing an uncurved seed. Campylotropous ovules have a curved embryo sac often giving the seed a tight "c" shape. The last ovule shape is called amphitropous, where the ovule is partly inverted and turned back 90 degrees on its stalk or funicle.

In the majority of flowering plants the zygotes first division is transversely orientated in regards to the long axis and this establishes the polarity of the embryo. The upper or chalazal pole becomes the main area of growth of the embryo, while the lower or micropylar pole produces the stalk-like suspensor that attaches to the micropyle. The suspensor absorbs and manufacturers nutrients from the endosperm that are utilized during the embryos growth.

The embryo is composed of different parts; the epicotyle will grow into the shoot, the radicle grows into the primary root, the hypocotyl connects the epicotyle and the radicle, the cotyledons form the seed leaves, the testa or seed coat forms the outer covering of the seed. Monocotyledonous plants like corn, have other structures; instead of the hypocotyle-epicotyle, it has a coleoptile that forms the first leaf and connects to the coleorhiza that connects to the primary root and adventitious roots form from the sides.

The seeds of corn are constructed with these structures; pericarp, scutellum (single large cotyledon) that absorbs nutrients from the endosperm, endosperm, plumule, radicle, coleoptile and coleorhiza - these last two structures are sheath-like and enclose the plumule and radicle, acting as a protective covering. The testa or seed coats of both monocots and dicots are often marked with patterns and textured markings, or have wings or tufts of hair.

Seed size and Seed Set

Seeds are very diverse in size. The dust-like orchid seeds are the smallest with about one million seeds per gram. Embryotic seeds have immature embryos and no significant energy reserves. They are myco-heterotrophs, depending on mycorrhizal fungi for nutrition during germination and the early growth of the seedling, in fact some terrestrial Orchid seedlings spend the first few years of their life deriving energy from the fungus and do not produce green leaves. At over 20 kg, the largest seed is the coco de mer.

Plants that produce smaller seeds can generate many more seeds while plants with larger seeds invest more resources into those seeds and normally produce fewer seeds. Small seeds are quicker to ripen and can be dispersed sooner, so fall blooming plants often have small seeds. Many annual plants produce great quantities of smaller seeds; this helps to ensure that at least a few will end in a favourable place for growth. Herbaceous perennials and woody plants often have larger seeds, they can produce seeds over many years, and larger seeds have more energy reserves for germination and seedling growth and produce larger, more established seedlings.

SEED FUNCTIONS

Seeds serve several functions for the plants that produce them. Key among these functions are nourishment of the embryo, dispersal to a new location, and dormancy during unfavourable conditions. Seeds fundamentally are a means of reproduction and most seeds are the product of sexual reproduction which produces a remixing of genetic material and phenotype variability that natural selection acts on.

Embryo Nourishment

Seeds protect and nourish the embryo or baby plant. Seeds usually give a seedling a faster start than a sporling from a spore gets because of the larger food reserves in the seed.

Seed Dispersal

Unlike animals, plants are limited in their ability to seek out favourable conditions for life and growth. As a consequence, plants have evolved many ways to disperse their offspring by dispersing their seeds. A seed must somehow "arrive" at a location and be there at a time favourable for germination and growth.

When the fruits open and release their seeds in a regular way, it is called dehiscent, which is often distinctive for related groups of plants, these fruits include; Capsules, follicles, legumes, silicles and siliques. When fruits do not open and release their seeds in a regular fashion they are called indehiscent, which include these fruits; Achenes, caryopsis, nuts, samaras, and utricles.

Seed dispersal is seen most obviously in fruits; however many seeds aid in their own dispersal. Some kinds of seeds are dispersed while still inside a fruit or cone, which later opens or disintegrates to release the seeds. Other seeds are expelled or released from the fruit prior to dispersal. Milkweeds produce a fruit type, known as a *follicle*, that splits open along one side to release the seeds. Iris capsules split into three "valves" to release their seeds.

By Wind (Anemochory)

Dandelion seeds (achenes) can be carried long distances by the wind.

- Many seeds (e.g. maple, pine) have a wing that aids in wind dispersal.
- The dustlike seeds of orchids are carried efficiently by the wind.
- Some seeds, (e.g. dandelion, milkweed, poplar) have hairs that aid in wind dispersal.

By Water (hydrochory)

- Some plants, such as *Mucuna* and *Dioclea*, produce buoyant seeds termed sea-beans or drift seeds because they float in rivers to the oceans and wash up on beaches.

By Animals (Zoochory)

- Seeds (burrs) with barbs or hooks (e.g. acaena, burdock, dock which attach to animal fur or feathers, and then drop off later.
- Seeds with a fleshy covering (e.g. apple, cherry, juniper) are eaten by animals (birds, mammals) which then disperse these seeds in their droppings.
- Seeds (nuts) which are an attractive long-term storable food resource for animals (e.g. acorns, hazelnut, walnut); the seeds are stored some distance from the parent plant, and some escape being eaten if the animal forgets them.

Myrmecochory

Myrmecochory is the dispersal of seeds by ants. Foraging ants disperse seeds which have appendages called elaiosomes (e.g. bloodroot, trilliums, Acacias, and many species of Proteaceae). Elaiosomes are soft, fleshy structures that contain nutrients for animals that eat them. The ants carry such seeds back to their nest, where the elaiosomes are eaten.

The remainder of the seed, which is hard and inedible to the ants, then germinates either within the nest or at a removal site where the seed has been discarded by the ants. This dispersal relationship is an example of mutualism, since the plants depend upon the ants to disperse seeds, while the ants depend upon the plants seeds for food. As a result, a drop in numbers of one partner can reduce success of the other.

In South Africa, the Argentine ant (*Linepithema humile*) has invaded and displaced native species of ants. Unlike the native ant species, Argentine ants do not collect the seeds of *Mimetes cucullatus* or eat the elaiosomes. In areas where these ants have invaded, the numbers of *Mimetes* seedlings have dropped.

Seed Dormancy and Protection

One important function of most seeds is delaying germination, which allows time for dispersal and prevents germination of all the seeds at one time when conditions appear favourable. The staggering of germination safeguards some seeds or seedlings from suffering during short periods of bad weather, transient herbivores or competition from other plants for light and nutrients. Many species of plants have seeds that germinate over many months or years, and some seeds can remain in the soil seed bank for more than 50 years before germination.

Seed dormancy is defined as a seed failing to germinate under environmental conditions optimal for germination, normally when the seed's environment is at the right temperature with proper soil moisture conditions. Induced dormancy or seed quiescence occurs when a seed fails to germinate because the external environmental conditions are inappropriate for germination, mostly in response to being too cold or hot, or too dry.

True dormancy or innate dormancy is caused by conditions within the seed that prevent germination under normally ideal conditions. Often seed dormancy is divided into four major categories exogenous; endogenous; combinational; and secondary. Exogenous dormancy is caused by conditions outside the embryo including

- Hard seed coats or physical dormancy occurs when seeds are impermeable to water or the exchange of gases. In some seeds the seed coat physically prevents the seedling from growing.
- Chemical dormancy includes growth regulators etc.

Endogenous dormancy is caused by conditions within the embryo itself, including

- Immature embryos where some plants release their seeds before the tissues of the embryos have fully differentiated, and the seeds ripen after they take in water while on the ground, germination can be delayed from a few weeks to a few months.
- Morphological dormancy where seeds have fully differentiated embryos that need to grow more before seed germination, the embryos are not yet fully developed.
- Morphophysiological dormancy seeds with underdeveloped embryos, and in addition have physiological components to dormancy. These seeds therefore require a dormancy-breaking treatments as well as a period of time to develop fully grown embryos.
- Physiological dormancy prevents seed germination until the chemical inhibitors are broken down or are no longer produced by the seed, often physiological dormancy is broken by a period of cool moist conditions, normally below (+4°C) 39°F, or in the case of many species in *Ranunculaceae* and a few others, (-5°C) 24°F. Other chemicals that prevent germination are washed out of the seeds by rainwater or snow melt. Abscisic acid is usually the growth inhibitor in seeds and its production can be affected by light. Some plants like Peony species have multiple types of physiological dormancy, one affects radical growth while the other affects shoot growth.
  - Drying; some plants including a number of grasses and those from seasonally arid regions need a period of drying before they will

germinate, the seeds are released but need to have a lower moister content before germination can begin. If the seeds remain moist after dispersal, germination can be delayed for many months or even years. Many herbaceous plants from temperate climate zones have physiological dormancy that disappears with drying of the seeds. Other species will germinate after dispersal only under very narrow temperature ranges, but as the seeds dry they are able to germinate over a wider temperature range.

– Photodormancy or light sensitivity affects germination of some seeds. These photoblastic seeds need a period of darkness or light to germinate. In species with thin seed coats, light may be able to penetrate into the dormant embryo. The presence of light or the absence of light may trigger the germination process, inhibiting germination in some seeds buried too deeply or in others not buried in the soil.

– Thermodormancy is seed sensitivity to heat or cold. Some seeds including cocklebur and amaranth germinate only at high temperatures (30°C or 86°F) many plants that have seed that germinate in early to mid summer have thermodormancy and germinate only when the soil temperature is warm. Other seeds need cool soils to germinate, while others like celery are inhibited when soil temperatures are too warm. Often thermodormancy requirements disappear as the seed ages or dries.

Combinational dormancy also called double dormancy. Many seeds have more than one type of dormancy, some *Iris* species have both hard impermeable seeds coats and physiological dormancy.

Secondary dormancy is caused by conditions after the seed has been dispersed and occurs in some seeds when none dormant seed is exposed to conditions that are not favourable to germination, very often high temperatures. The mechanisms of secondary dormancy is not yet fully understood but might involve the lose of sensitivity in receptors in the plasma membrane.

Many garden plants have seeds that will germinate readily as soon as they have water and are warm enough, though their wild ancestors may have had dormancy, these cultivated plants lack seed dormancy. After many generations of selective pressure by plant breeders and gardeners dormancy has been selected out.

For annuals, seeds are a way for the species to survive dry or cold seasons. Ephemeral plants are usually annuals that can go from seed to seed in as few as six weeks.

Not all seeds undergo a period of dormancy. Seeds of some mangroves are viviparous, they begin to germinate while still attached to the parent. The large, heavy root allows the seed to penetrate into the ground when it falls.

SEED STRUCTURE

A seed contains the embryo from which a new plant will grow under proper conditions. Seeds also usually contain a supply of stored energy and is wrapped in the seed coat or testa. Seeds are very diverse in size. The dust-like orchid seeds are the smallest with about one million seeds per gram. Embryotic seeds have immature embryos and no significant energy reserves. They are myco-heterotrophs, depending on mycorrhizal fungi for nutrition during germination. At over 20 kg, the largest seed is the coco de mer.

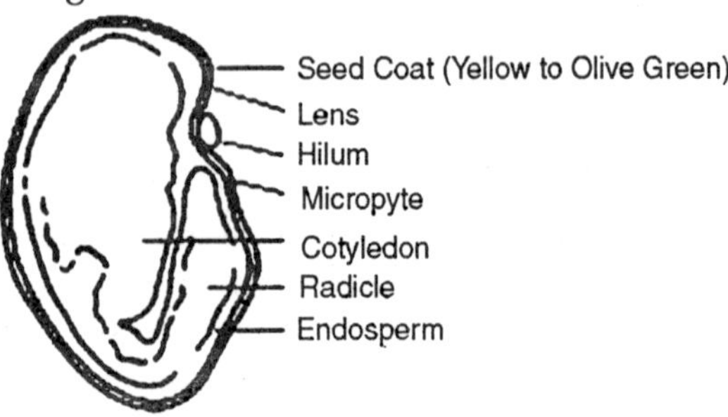

Fig. Alfalfa Seed Structure

The embryo has one cotyledon or seed leaf in monocotyledons, two cotyledons in almost all dicotyledons and two or more in gymnosperms. The radicle is the embryonic root. The plumule is the embryonic shoot. The embryonic stem above the point of attachment of the cotyledon(s) is the epicotyl. The embryonic stem below the point of attachment is the hypocotyl.

In angiosperms, the stored food begins as a tissue called the endosperm, which is derived from the parent plant via double fertilization. The usually triploid endosperm is rich in oil or starch and protein. In gymnosperms, such as conifers, the food storage tissue is part of the female gametophyte, a haploid tissue. In some species, the embryo is embedded in the endosperm or female gametophyte, which the seedling will use upon germination.

In others, the endosperm is absorbed by the embryo as the latter grows within the developing seed, and the cotyledons of the embryo become filled with this stored food. At maturity, seeds of these species have no endosperm and are termed exalbuminous seeds. Some exalbuminous seeds are bean, pea, oak, walnut, squash, sunflower, and radish. Seeds with an endosperm at maturity are termed albuminous seeds. Most monocots (e.g. grasses and palms) and many dicots (e.g. brazil nut and castor bean) have albuminous seeds. All gymnosperm seeds are albuminous. The seed coat develops from the tissue, the integument, originally surrounding the ovule.

The seeds of angiosperms are contained in a hard or fleshy (or with layers of both) structure called a fruit. Gymnosperm seeds begin their development "naked" on the bracts of cones, although the seeds do become covered by the cone scales as they develop. An example of a hard fruit layer surrounding the actual seed is that of the so-called *stone* fruits (such as the peach). Some seeds have an appendage on the seed coat such an aril (as in yew and nutmeg) or an elaiosome (as in Corydalis) or hairs (as in cotton). The hilum is the scar on the seed coat where the seed was attached to the ovary wall by the funiculus.

In order for the seed coat to split, the embryo must imbibe

(soak up water), which causes it to swell, splitting the seed coat. However, the nature of the seed coat determines how rapidly water can penetrate and subsequently initiate germination. For seeds with a very thick coat, scarification of the seed coat may be necessary before water can reach the embryo.

Examples of scarification include gnawing by animals, freezing and thawing, battering on rocks in a stream bed, or passing through an animal's digestive tract. In the latter case, the seed coat protects the seed from digestion, while perhaps weakening the seed coat such that the embryo is ready to sprout when it gets deposited (along with a bit of fertilizer) far from the parent plant. In species with thin seed coats, light may be able to penetrate into the dormant embryo.

The presence of light or the absence of light may trigger the germination process, inhibiting germination in some seeds buried too deeply or in others not buried in the soil. Abscisic acid is usually the growth inhibitor in seeds. A fertilized seed contains the embryo from which a new plant will grow under proper conditions. It also contains a supply of stored food and is wrapped in the seed coat or testa. The stored food begins as a tissue called endosperm derived from the parent plant. Endosperm becomes rich in oil or starch, and protein. In some species, the embryo is imbedded in the endosperm, which the seedling will use upon germination. In others, the endosperm is absorbed by the embryo as the latter grows within the developing seed.

A fertilized seed contains the embryo from which a new plant will grow under proper conditions. It also contains a supply of stored food and is wrapped in the seed coat or *testa*. The stored food begins as a tissue called *endosperm* derived from the parent plant. Endosperm becomes rich in oil or starch, and protein.

In some species, the embryo is imbedded in the *endosperm*, which the seedling will use upon germination. In others, the endosperm is absorbed by the embryo as the latter grows within the developing seed, and the cotyledons of the embryo become filled with this stored food. At maturity, seeds of these species have no endosperm. Some common plant seeds that lack an endosperm are bean, pea, oak, walnut, squash, sunflower, and radish. Plant seeds with an endosperm include all conifers and most monocotyledons (e.g. grasses and palms), and also e.g. brazil nut, castor bean.

The seed coat develops from tissues (called *integument*) originally surrounding the ovule. The seed coat in the mature seed can be a paper thin layer or something more substantial. The seed coat helps protect the embryo from mechanical injury and from drying out. In order for the seed coat to split, the embryo must imbibe (soak up water) which causes it to swell, splitting the seed coat.

However, the nature of the seed coat determines how rapidly water can penetrate and subsequently initiate germination. For seeds with a very thick coat, scarification of the seed coat may be necessary before water can reach the embryo. Examples of scarification include gnawing by animals, freezing and thawing, battering on rocks in a stream bed, or passing through an animal's digestive tract.

In the latter case, the seed coat protects the seed from digestion, while perhaps weakening the seed coat such that the embryo is ready to sprout when it gets deposited (along with a bit of fertilizer) far from the parent plant. In species with thin seed coats, light may be able to penetrate into the dormant embryo. The presence of light or the absence of light may trigger the germination process, inhibiting germination in some seeds buried too deeply or in others not buried in the soil. Abscisic acid is usually the growth inhibitor in seeds.

The seeds of angiosperms are contained in a hard or fleshy (or with layers of both) structure called a fruit. Gymnosperm seeds begin their development "naked" on the bracts of cones, although the seeds do become covered by the cone scales as they develop. An example of a hard fruit layer surrounding the actual seed is that of the so-called *stone* fruits (such as the peach).

Alfalfa is a dicotyledonous plant. That is, the seed is composed of two embryonic or seed leaves called cotyledons. In addition, the seed contains the primary root or radicle, the shoot growing point or epicotyl located above the cotyledons, and the endosperm or food storage area. Externally the seed has several visible structures – the hilum or point of attachment in the seed pod, the lens or a weak point in the seed coat, and the micropyle, which is a remnant of the tiny opening the pollen tube grew through during the process of fertilizing the female ovary.

ALFALFA SEED STRUCTURE

Germination is the resumption of growth by the embryo within the seed leading to the development of a new plant. The germination process is complete as soon as the radicle ruptures the seed coat.

The process is influenced by available soil moisture, soil temperature, and the nature of the soil surrounding the seed (i.e., salty vs. normal soil) or residual herbicides that may be present in the soil.

ALFALFA GERMINATION PROCESS

Seedling growth is the developmental period of the young plant from the time germination is completed until it can manufacture enough food through photosynthesis to sustain growth. The seedling root is the first structure to emerge from the seed during germination. It penetrates the soil very rapidly, forming a slender, usually unbranched taproot, which may penetrate 5 to 6 feet into the soil during the first growing season.

Once the seedling root is anchored firmly in the soil, the seedling axis below the cotyledons elongates in an arch (hypocotyl arch) pulling the cotyledons upward to the soil surface. Seed germination and seedling emergence occur in about three to seven days. Fewer days may be required for seedling emergence under ideal soil moisture and temperature conditions. As the hypocotyl arch emerges from the soil, growth stops on the side exposed to light and continues on the underside until the seedling is in an upright position.

This raises the cotyledons above the soil surface where they expand. The growing point (epicotyl) of the seedling is now exposed. The first true leaf is unifoliolate and emerges from a bud at the first stem node above the cotyledons. Seedling growth is complete. Under good growing conditions, the seedling is developed fully 10 to 15 days after planting.

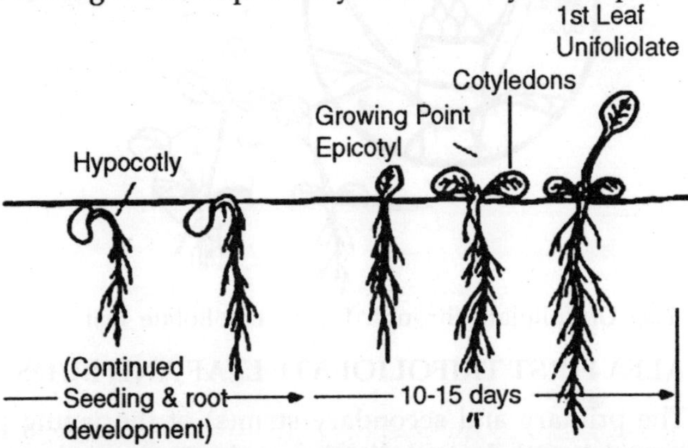

Fig. Seedling Emergence Through Unifoliolate Leaf Stage

ALFALFA SEEDLING GROWTH

A fully developed alfalfa seedling does not assure plant establishment. It must continue to develop deeper roots and grow more leaves to survive and become an established stand. Vegetative growth continues through cell division and expansion in the epicotyl or growing point of the young plant.

The second leaf of the alfalfa plant is usually trifoliolate (three leaflets) and originates from the second primary stem node. All subsequent leaves are trifoliolates except in new multifoliolate varieties that have 5, 7, or 9 leaflets per leaf. The unifoliolate leaf and all subsequent trifoliolate/multifoliolate leaves are useful characters to distinguish alfalfa from weed seedlings in stand evaluation.

Axillary buds develop in the axils of all leaves. After three or more trifoliolate/multifoliolate leaves have appeared on the primary stem, new secondary stem growth may occur from any of the axillary buds, but frequently only one axillary bud stem (usually the unifoliolate) develops early, especially if seedlings are shaded in a companion crop or under competition from weed growth or other alfalfa plants.

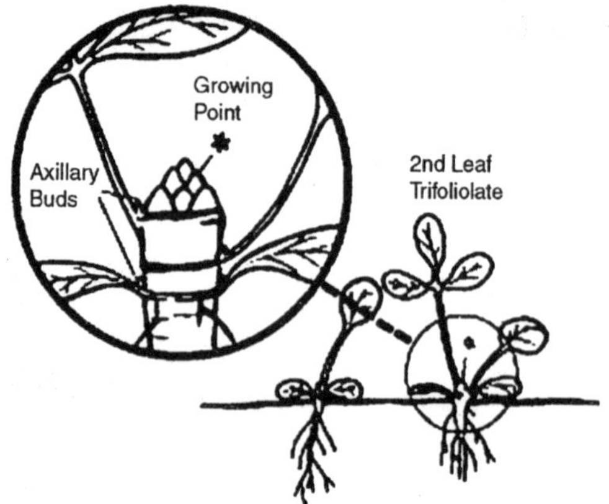

Fig. Unifoliolate through fourth trifoliolate leaf stage

ALFALFA FIRST TRIFOLIOLATE LEAF AND BUDS

The primary and secondary stem(s) of the young plant increase in length by cell division and internode elongation from the first stem node upwards. The second and subsequent leaves are trifoliolate/multifoliolate and develop alternately at each stem node as growth continues.

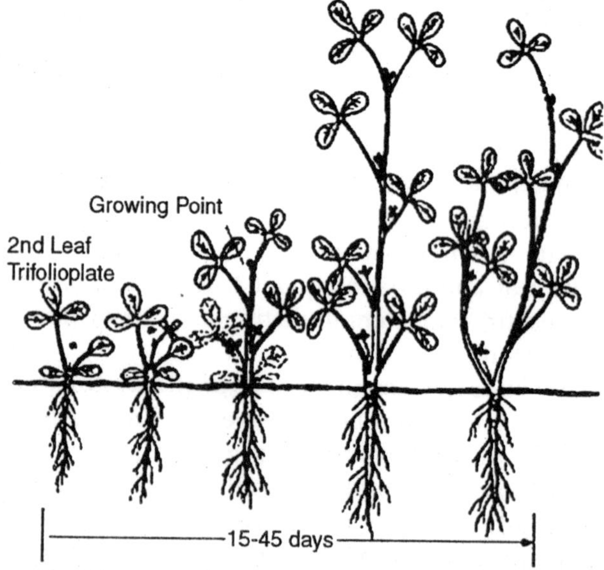

Fig. First-year Vegetative Growth and Development

Once the first true trifoliolate leaf develops, further growth and development is best described by the number of trifoliolate/multifoliolate leaves that develop on the main shoot as the plant continues to grow. Any axillary buds in axils of leaves can develop new stem tissue and generally do in less competitive conditions.

ALFALFA FIRST-YEAR VEGETATIVE DEVELOPMENT

Growth and development of new shoots from axillary buds gives the young plant a branched appearance, especially if light is adequate and the stand is not too thick. Vigourously growing alfalfa plants quite often have three and sometimes four secondary shoots in addition to the primary or central stem, which forms the characteristic first-year crown.

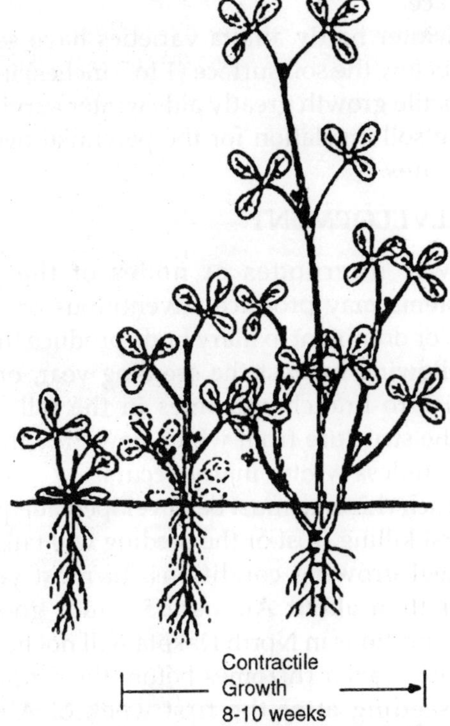

Fig. Crown Development Showing Contractile Growth

The branches from the unifoliolate leaf bud appears first,

followed by branches from the cotyledonary and first trifoliolate leaf buds. Fewer branches may form the crown with late seedings, under severe companion crop competition, or with high seeding rates. Under these conditions the cotyledonary, unifoliolate, and first trifoliolate leaf buds may remain dormant or may give rise to a branch rhizome or underground stem in the fall.

The young alfalfa plant, at about 8 to 10 weeks of age, undergoes a growth phase known as contractile growth. This process in alfalfa and sweetclover involves a change in the shape of cells in the hypocotyl or seedling axis below the cotyledons and upper portion of the primary root from long and narrow to short and wide as a result of carbohydrate or food storage. This shift pulls the lower stem nodes beneath the soil surface.

Most winter hardy alfalfa varieties have several nodes pulled well below the soil surface (1 to 3 inches) in the seeding year. Contractile growth greatly aids winter survival of alfalfa by providing soil insulation for the perennial over-wintering crown structures.

CROWN DEVELOPMENT

The lower internodes or nodes of the primary or secondary stems may produce adventitious or crown buds. Crown buds or dominant axillary buds produce the vegetative regrowth following harvest the seedling year, or these buds may give rise to branch rhizomes in the fall. The branch rhizome is the structure from which new growth will initiate in the spring unless winter injury occurs.

The branch rhizome must be developed adequately in the fall by the first killing frost or the seeding will fail, even under the most ideal growing conditions. In most years, alfalfa seeded later than about August 15 under good moisture (irrigation) conditions in North Dakota will not have adequate time to initiate branch rhizomes before the first killing frost. A dryland seeding after the first week of August is not recommended because of the increased risk of stand loss.

A crown of a perennial can be described best as the young

overwintering stem tissue. The crown in the fall of the first year may be as small as the lower portion of the main stem and dormant cotyledonary node buds, or as large as the lower portion of the main stem (i.e., colytedonary, unifoliolate, and first trifoliolate leaf nodes), the secondary stems from these nodes, and the branch rhizomes that develop.

Spring growth will occur from the rudimentary leaves on the branch rhizome if winter injury hasn't "burnt" them off. If winter injury has occurred, spring growth will initiate from adventitious buds on the branch rhizome or from a dormant crown bud. But, spring growth is much slower in this case. All subsequent growth in the first-harvest year occurs from adventitious or dormant crown buds.

The crown of the alfalfa plant increases in size during the second year. Branch rhizomes develop on last year's branch rhizome. These grow outward and upward increasing the circumference of the plant. After two to four years, the typical multibranched crown of alfalfa develops.

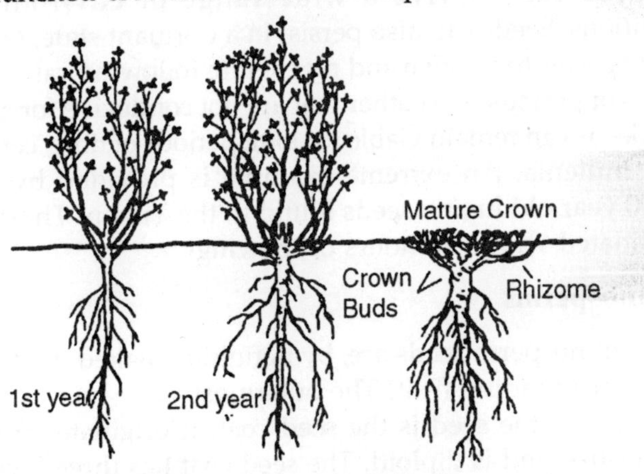

Fig. Fully Developed Alfalfa Crown

ALFALFA WINTERING STRUCTURES

The crown and associated root tissue are the storage centre for carbohydrates (food) produced through the process of photosynthesis. The carbohydrates are used to develop cold resistance or winter hardiness, for plant respiration during winter, and to provide the energy source for the initiation of the new growth in the spring and following each harvest. First-year management influences subsequent growth. Alfalfa management is discussed in R-571, Alfalfa Management in North Dakota.

Over the next several lectures we will follow the early growth of young trees and other vascular plants from seed through germination and growth in the first year. We will draw examples from many species, but will illustrate some phases of growth by contrasting two tree species native to British Columbia Douglas-fir representing gymnosperms, and red alder representing angiosperms.

Evolution of Seeds

The evolution of seeds was a critical factor in facilitating the proliferation of land plants into a wide range of environments. Seeds protect and nourish embryos, and embryos can survive a wide range of environmental conditions. Seeds can also persist in a dormant state, enabling plant species to survive and regenerate following catastrophic events or periods of weather that are not conducive for growth.

Seeds can remain viable for long periods of time, centuries, even millenia. An extreme example is provided by frozen 10,000 year old lupine seeds found in the Yukon. These seeds germinated within 48 hours of thawing!

Gymnosperms

Gymnosperm seeds are, by definition, naked, that is they lack a surrounding fruit. The outermost

layer of the seed is the seed coat. It originates from the mother tree and is diploid. The seed coat has three layers the outer layer; the thicker, tough 'stony' or middle layer; and the inner layer. The colour of the seed coat, which can be quite variable, is due to tannins that accumulate during seed development. Some species of conifers, including *Abies* spp. (true firs), *Tsuga* spp. (hemlocks) and *Thuja plicata* (western redcedar) have resin vesicles in the middle or outer layers of the seed coat.

These resin vesicles may play a role in:

- Seed coat dormancy,
- Protecting the embryo from excessive dehydration,
- Deterring seed herbivory.

These vesicles can also create problems for the extraction and cleaning of conifer seed. Many conifer seeds have a seed wing to aid in dispersal.

Immediately inside the cell wall is the nucellus, a papery layer surrounding the megaspore cell wall. Inside the megaspore cell wall is the megagametophyte (the female gametophyte), the haploid nutritional tissue found in gymnosperm seeds. Early in the development of cones, megagametophytes produce egg cells which are fertilized by male gametes to produce zygotes. Zygotes develop into embryos. The megagametophyte then plays its second functional role, surrounding the embryo, protecting and nourishing it.

The megagametophyte in Douglas-fir is 60% lipids, 16% proteins, and 2% sugars, making it a high-energy and nutrient tissue both for the embryos it contains and for a plethora of seed predators including small mammals, birds and insects. In a dry seed (<10% moisture), the megagametophyte is creamy yellow and does not entirely fill the seed coat. When a seed is imbibed (soaked in water until the moisture content is 30-40%), the megagametophyte is white and shiny.

The embryo is found in the corrosion cavity, a pit in the centre of the megagametophyte that is fully filled by the embryo in mature seeds. It consists of the cotyledons, shoot apical meristem, root apical meristem, root cap and suspensor. The cotyledons and shoot apical meristem point towards the wider chalazal end of the seed.

The radicle (embryonic root including the apical meristem and root cap) and suspensor are at the more pointed micropylar end of the seed towards the micropyle (the opening in the integument through which the pollen entered the ovule). The number of cotyledons varies from 2 in western redcedar to 12 or more in *Pinus ponderosa* (ponderosa pine).

The suspensor is found at the base of the root cap and

plays a role early in embryo development by pushing the embryo into the megagametophyte.

Angiosperms

Angiosperm seeds are found inside fruit. The many types of fruit will be discussed in detail in lab. Angiosperm seeds generally vary more in structure than conifer seeds, but share some features with conifer seeds and have different parts playing the same functional roles for others.

Like conifer seeds, angiosperm seeds are surrounded by a seed coat protecting the embryo. Inside the seed coat (which is usually formed by the integument, but is sometimes fused with the fruit), nutritive reserves for the mature embryo may be stored either in endosperm or in large cotyledons.

Endosperm is formed through the fertilization of the two polar nuclei by a male gamete, producing triploid cells, at the same time that the egg cell is fertilized by a second male gamete (more on this after the Christmas break). All angiosperm seed have endosperm early in embryo development. In mature seeds of some species, there may be a substantial amount of triploid endosperm surrounding the embryo (e.g., *Arbutus menziesii* (arbutus)).

Commonly in Eudicots, however, embryos largely absorb the endosperm and the nutritional reserves are stored in the two (of course!) large cotyledons that fill the seed cavity (e.g., *Alnus* spp. (alders), *Acer* spp. (maples), and *Quercus* spp. (oaks)). Some angiosperm seeds already contain a primordial shoot with leaves, called a plumule, above (distal to) the cotyledons prior to germination.

The seeds of monocots have several noteworthy features. They typically have endosperm at maturity, and the single cotyledon often plays a role in absorbing food digested from the endosperm. Grass embryos have a large cotyledon called a scutellum attached to one side, as well as a plumule and a radicle. The plumule is protected by a structure called the coleoptile, while the radicle is protected by the coleorhiza.

Angiosperm seed also have suspensors, but these have a different function than those of conifers. In angiosperms, suspensors are involved in supporting the early development of the embryo both through providing nutrition and growth regulators. The suspensor cells die in angiosperms relatively early in embryo development.

SEED DISPERSAL

Conifers

The dispersal of winged seeds depends on seed mass, wing size and wind speed. Douglas-fir seeds have a mass of 9 to 17 mg, are 1 to 1.5 cm in length including wings, and can disperse up to 3 or 4 tree heights from the mother tree. Periodic weather events may result in rare dispersals up to a km in distance.

Most BC conifers shed seed August through October. Cone tissues die, cones dry out, cone scales flex and release seed. Dry conditions promote the flexing of cone scales and opening of cones, so cycles of wet and dry weather may produce repeated cycles of seed release. Seeds that land on snow may blow considerable distances along the frozen surface in winter.

Not all conifers have cones that open on maturity. The true firs (*Abies* spp.) bear upright rather than pendant cones, and have deciduous cone scales. When these cones are mature, the cone scales fall off the central cone rachis, releasing the seed. Whitebark pine (*Pinus albicaulis*) cones do not open and must be mechanically torn apart to release seeds. This is done by Clark's nutcrackers, which then disperse store seeds for later consumption (and regeneration of the pine).

Angiosperms

Dispersal of seeds in angiosperms can be via wind, through animal ingestion or through animal transport to food caches or as 'hitchhikers' (e.g., burrs). The acorns of oaks, and the many fleshy fruits and tasty nuts produced by angiosperms, provide examples of angiosperms dispersed by animals.

Red alder is a wind-dispersed species with single seeds

found in small (1.5 million per kg), winged nutlets. The fruit are shed in large numbers in mid to late fall, and can be wind-dispersed in air or across snow. Red alder is a good example of a pioneer species, capable of colonizing bare ground. Pioneer species typically have small seed produced in large numbers that disperse widely via wind.

Species of poplars and aspen (*Populus* spp.) are also pioneers, and produce large quantities of tiny seed (8 million per kg for trembling aspen). Poplar seed contain very small nutritional reserves and thus are quite short-lived. Fireweed (*Epilobium angustifolium*) is a good example of an herbaceous pioneer with seed production and dispersal similar to poplars.

SEED DORMANCY

Seed of most temperate and boreal species is dormant when it disperses, that is it won't germinate even if conditions of temperature, moisture and light are favourable. Why? To avoid active growth until winter has passed.

An actively growing plant cannot be adequately frost hardy. Dormant seeds cease growth and maintain a very low metabolic rate until specific environmental signals are received or other requirements have been met.

Embryo Dormancy

Dormancy in many species is under the physiological control of the embryo. Embryo dormancy results from the levels of plant growth regulators (plant hormones). Dormant embryos have growth-inhibitors present and lack growth promoters. Breaking embryo dormancy is dependent on decreasing growth inhibitor concentrations and increasing growth promoters above some threshold.

Embryo dormancy in many species can be overcome through exposure to low temperatures to meet the chilling requirement of the seed. To accomplish this, seeds are imbibed (soaked in water), then placed in a fridge, usually on damp paper or moss. This is called stratification.

In Douglas-fir, holding seeds at 4°C for 4 weeks will meet the chilling requirement. Many species require a longer stratification period – up to 100 days. Seeds of some species with embryo dormancy may germinate eventually without chilling after being soaked in water, but will germinate more slowly than stratified seed, and the percent of seeds that germinate will likely be lower.

Embryo dormancy can also be overcome in some cases by artificially applying growth regulators. Some species require alternating day and night temperatures, or cycles of wetting and drying following low temperatures. Some species have both embryo dormancy and other types of seed dormancy.

Seed Coat Dormancy

Some desert and tropical species have chemical growth inhibitors in the seed coat that must be

leached out by water before seed will germinate, increasing the probability that seed will germinate during adequately moist periods. Other species require a heat treatment to germinate. Germination of buckbrush (*Ceonothus* spp.) is promoted by pouring boiling water over seeds.

OTHER TYPES OF SEED DORMANCY

Embryo immaturity – Seeds of some species are immature at the time fruit or seed are shed and require 'after-ripening'. The seed of whitebark pine are typically immature when harvested by Clark's nutcrackers and continue to mature after caching.

Seed coat impermeability – Seed coats can be impermeable to water or oxygen, which physically imposes dormancy through preventing growth. This is common for leguminous species, among others. These seed may need scarification (scratching or abrading of the seed coat, either in nature or artificially with sandpaper), infection of seed coat by fungi or bacteria, or treatment with concentrated sulphuric acid (simulating passing the seed through the digestive tract of an animal).

Light requirement – Some species (e.g, some spruce and pine) require exposure to light to germinate. Other species require exposure to light of a specific quality. Red light promotes germination in these species, while far red light inhibits it. It is the ratio of far red to red light that is critical. As light passes through vegetation, this ratio increases, and germination in light-sensitive species will decrease. This means that seeds will be less likely to germinate when they are below competing vegetation. Once the competing vegetation is gone, the amount of red to far red light will increase, and the seed will germinate.

Serotiny – Most interior lodgepole pine (*Pinus contorta* ssp. *latifolia*) has cones that are serotinous. The cone scales are 'glued' closed by resin that has a melting point around 40°C. These cones must be exposed to either a fire or to particularly warm temperatures (e.g., on a dark surface in the full sun) to open. The cones of juvenile interior lodgepole pine, and Shore pine (*P. contorta* ssp. *contorta*) do not have this adaptation – why?

Species Without Seed Dormancy

Red alder seeds lack dormancy, but since they are shed late in the fall, environmental conditions are not favourable for germination until spring. Other examples of seeds with no dormancy include sugar maple and willows, which are viable for only a few days following dispersal.

Many agricultural crop species have no seed dormancy, but their wild progenitors do – humans have selected against seed dormancy in these species as it complicates farming.

GERMINATION

Germination is the resumption of growth by embryos when exposed to favourable conditions. It requires several steps

- Seeds imbibe and swell; the seed coat ruptures;
- Cell division occurs in the embryo and the radicle elongates and extends out of the seed;
- Further elongation of the radicle pushes either the cotyledons or just the epicotyl hormones produced in the meristems and possibly elsewhere initiate enzymatic activity, hydrolysis of stored nutrients, and

movement of these nutrients to regions of growth.

The process of germination is complete when enough photosynthetic tissue is exposed to support further growth. The radicle develops into the tap root of the seedling (except in the case of species with fibrous root systems). It is positively gravitropic (grows in the direction of gravitational pull) while the shoot is negatively gravitropic (grows against gravity). The root cap consists of loose, slimy cells that protect the root apical meristem as the root elongates through the soil. The radicle develops root hairs for absorption of water and dissolved nutrients soon after germination.

In species with epigeous germination (e.g., conifers, red alder, maples), the cotyledons are pushed out of the ground, turn green and become photosynthetically active. In species with hypogeous germination (e.g., oaks and other species with large seeds), the cotyledons remain in the ground and only the epicotyl emerges. The energy reserves in the cotyledons supply the energy for early seedling growth, but the cotyledons never become photosynthetically active in this case. Rather, the first true leaves are the first photosynthetic organs with hypogeous germination.

Symbiotic relationships between seedling roots and other organismsareofteninitiatedsoonaftergermination.Lateralroots canbecomeinfectedbymycorrhizalfungisoonaftergermination. Inadditiontobecomingmycorhizal,redaldergerminantsbecomes symbiotic with *Frankia* bacteria. Nodules form on the roots, and in these nodules N_2 gas is fixed into NH_4^+, which can then be convertedintoaminoacids.Thisallowsredalderto'fertilizeitself'. Up to 320 kg N can be fixed per ha per year! Other species that fix nitrogen include *Ceonothus* spp. (deer brush) and *Shepherdia canadensis* (soopallalie).

SEED DEVELOPMENT

Following the fertilizations in the embryo sac, the zygote divides repeatedly by mitosis and differentiates into an embryo. The endosperm nucleus also divides by mitosis and forms the endosperm tissue, which provides food for the developing embryo.

The early embryo is linear with apical meristems on either end and one or two seed leaves or cotyledons. The axis below the cotyledons is called the hypocotyl, at the tip of which is the radicle that gives rise to the primary root of the seedling. The axis above the attachment of the cotyledons is the epicotyl, which also ends in an apical meristem. In some seeds, the first foliage leaves are formed in the seed.

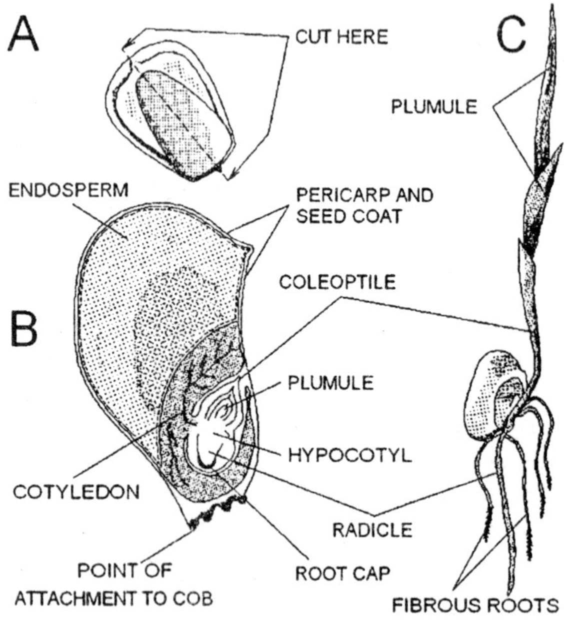

Fig. Seed Structure Monocot.

The area above the cotyledons is thus a miniature shoot and is called the plumule. In some taxa, food for the embryo remains within endosperm tissue and the cotyledons serve as organs of absorption. In others, food moves directly into the embryo and is stored within the cotyledons, leaving only a minuscule endosperm. In still another variation, the nucellus (the megasporangium wall) enlarges and becomes a storage tissue called perisperm. The integuments harden into the seed coat as the embryo matures. The scar left on the seed coat by the separation of the funiculus from the integuments is called the hilum. Often the micropyle remains visible near the hilum.

Angiosperms traditionally have been separated into two major categories on the basis of the number of cotyledons they possess monocots (mono = one; cotyledons = seed leaves) and dicots (di = two; cotyledons = seed leaves). Monocots are the grasses, sedges, lilies, and their relatives, while dicots constitute the rest of the flowering plants.

This artificial separation currently is being replaced by a more natural classification in which the *monocots* are retained as a natural group, but the dicots are separated into the eudicots (eu = true, dicots) and the magnoliids. The latter is a small group of very primitive angiosperms ancestral to both monocots and eudicots and has both woody representatives—woody magnoliids (trees like *Magnolias,* tulip trees, and laurels)—and paleoherbs (herbaceous plants such as members of the pepper and water lily families). The eudicots constitute about 97 percent of the angiosperms, while the magnoliids make up the other 3 percent.

In monocot seeds the single cotyledon usually digests and absorbs food from the endosperm and translocates it to the embryo. In grasses, the cotyledon is called the scutellum. In grasses, also, there is a protective sheath called the coleoptile over the plumule and another, the coleorhiza, surrounding the radicle.

SEED GERMINATION

Internal signals shut off growth of the embryo at a certain size and the seed goes into a period of dormancy. In some plants, dormancy lasts only as long as it takes the seed to be dispersed from the ovary. In others, dormancy may last for long periods until either external or internal signals (or a combination of both) initiate further growth. Environmental factors of chief importance to initiate growth are water, light, and temperature.

SEEDLING GROWTH

When an embryo resumes growth, stored food provides the energy for seedling development—the roots first, followed by elongation of the photosynthetic shoots.

Eudicot Development

Aboveground growth in eudicots takes one of two general patterns epigeous or hypogeous. In epigeous growth, the hypocotyl elongates, pulling the plumule and cotyledons above ground; in hypogeous growth, the cotyledons remain below ground because the epicotyl grows faster than the hypocotyl and pulls the plumule erect.

Monocot Development

Monocots develop with two different general patterns one for the grasses, one for the rest of the group. In most of the monocots (but not grasses), after the radicle has pushed out of the seed coat, the first shoot structure to emerge is the cotyledon, which arches upward with the remainder of the endosperm and the seed coat still attached. It elongates above ground and is photosynthetic until the true leaves develop.

In the grasses, the sheaths around both the shoot and the root tip must be penetrated by the roots and the shoots. The root sheath, the coleorhiza, grows faster than the radicle for a short time, but when it stops growing, the radicle emerges and forms an anchoring primary root. The shoot sheath, the coleoptile, moves upward to the soil surface through elongation of the first internode of the stem (called the *mesocotyl*), and when it reaches the surface it stops growing.

The plumule then pushes through into the air. At about the same time, buds of adventitious roots begin to grow and, by the time the seedling is erect with a few true leaves, it already has adventitious roots growing downward from its first node. The primary root system is short lived and dies soon after its establishment and the adventitious root system becomes the principal absorbing and anchoring system for the new grass plant.

GERMINATION

Germination is the process whereby growth emerges from a period of dormancy. The most common example of germination is the sprouting of a seedling from a seed of an angiosperm or gymnosperm.

Fig. Sunflower Seedlings, Just three days after Germination

However, the growth of a sporeling from a spore, the growth of hyphae from fungal spores, is also germination. In a more general sense, germination can imply anything expanding into greater being from a small existence or germ.

Pollen Germination

Another germination event during the life cycle of gymnosperms and flowering plants is the germination of a pollen grain after pollination. Like seeds, pollen grains are severely dehydrated before being released to facilitate their dispersal from one plant to another. They consist of a protective coat containing several cells (up to 8 in gymnosperms, 2-3 in flowering plants). One of these cells is a tube cell.

Once the pollen grain lands on the stigma of a receptive flower (or a female cone in gymnosperms), it takes up water and germinates. Pollen germination is facilitated by hydration on the stigma, as well as the structure and physiology of the stigma and style. Pollen can also be induced to germinate *in vitro* (in a petri dish or test tube).

During germination, the tube cell elongates into a pollen tube. In the flower, the pollen tube then grows towards the ovule where it discharges the sperm produced in the pollen grain for fertilization. The germinated pollen grain with its two sperm cells is the mature male microgametophyte of these plants.

Self-incompatibility

Since most plants carry both male and female reproductive organs in their flowers, there is a high risk for self-pollination and thus inbreeding. Some plants use the control of pollen germination as a way to prevent this selfing. Germination and growth of the pollen tube involve molecular signaling between stigma and pollen. In self-incompatibility in plants, the stigma of certain plants can molecularly recognize pollen from the same plant and prevents it from germinating.

SPORE GERMINATION

Germination can also refer to the emergence of cells from resting spores and the growth of sporeling hyphae or thalli from spores in fungi, algae, and some plants.

Resting Spores

In resting spores, germination involves cracking the thick cell wall of the dormant spore. In zygomycetes the thick-walled zygosporangium cracks open and the zygospore inside gives rise to the emerging sporangiophore. In slime molds, germination refers to the emergence of amoeboid cells from the hardened spore. After cracking the spore coat, further development involves cell division, but not necessarily the development of a multicellular organism.

Zoospores

In motile zoospores, germination frequently means a lack of motility and changes in cell shape, which allow the organism to become sessile.

Ferns and Mosses

In plants such as bryophytes, ferns, and a few others, spores germinate into independent gametophytes. In the bryophytes (e.g. mosses and liverworts), spores germinate into protonemata, similar to fungal hyphae, from which the gametophyte grows. In ferns, the gametophytes are small, heart-shaped prothalli that can often be found underneath a spore-shedding adult plant.Seedling

A seedling is a young plant sporophyte developing out of a plant embryo from a seed. Seedling development starts with germination of the seed. A typical young seedling consists of three main parts the radicle (embryonic root), the hypocotyl (embryonic shoot), and the cotyledons (seed leaves). The two classes of flowering plants are distinguished by their numbers of seed leaves Monocotyledons (monocots) have one blade-shaped cotyledon, whereas dicotyledons (dicots) have two round cotyledons. Gymnosperms are more varied. Pine seedlings have up to eight cotyledons. The seedlings of some flowering plants have no cotyledons at all. These are said to be acotyledons.

Germination and Early Seedling Development

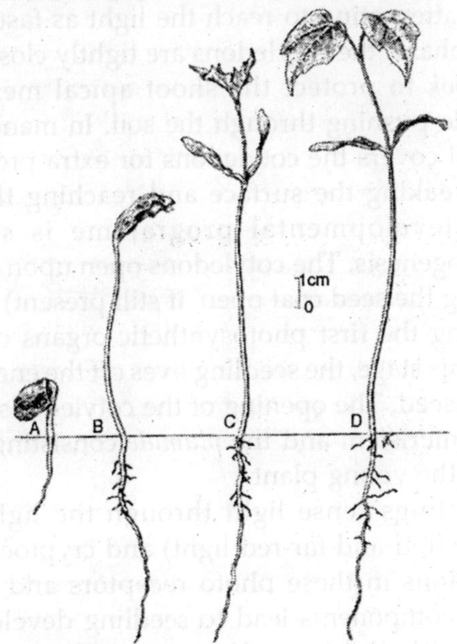

Fig. Development of an Angiosperm (maple) Seedling

During germination, the young plant emerges from its protective seed coat with its radicle first, followed by the cotyledons. The radicle orients towards gravity, while the hypocotyl orients away from gravity and elongates through cell expansion to push the cotyledons out of the ground.

PHOTOMORPHOGENESIS AND SKOTOMORPHOGENESIS

Dicot seedlings grown in the light develop short hypocotyls and open cotyledons exposing the epicotyl. This is also referred to as photomorphogenesis. In contrast, seedlings grown in the dark develop long hypocotyls and their cotyledons remain closed around the epicotyl in an *apical hook*. This is referred to as skotomorphogenesis or etiolation. Etiolated seedlings are yellowish in colour as chlorophyll synthesis and chloroplast development depend on light. They will open their cotyledons and turn green when treated with light.

In a natural situation, seedling development starts with skotomorphogenesis while the seedling is growing through the soil and attempting to reach the light as fast as possible. During this phase, the cotyledons are tightly closed and form the *apical hook* to protect the shoot apical meristem from damage while pushing through the soil. In many plants, the seed coat still covers the cotyledons for extra protection.

Upon breaking the surface and reaching the light, the seedling's developmental programme is switched to photomorphogenesis. The cotyledons open upon contact with light (splitting the seed coat open, if still present) and become green, forming the first photosynthetic organs of the young plant. Until this stage, the seedling lives off the energy reserves stored in the seed. The opening of the cotyledons exposes the shoot apical meristem and the *plumule* consisting of the first *true leaves* of the young plant.

The seedlings sense light through the light receptors phytochrome (red and far-red light) and cryptochrome (blue light). Mutations in these photo receptors and their signal transduction components lead to seedling development that is at odds with light conditions, seedlings that show photomorphogenesis when grown in the dark.

Seedling Growth and Maturation

Once the seedling starts to photosynthesize, it is no longer dependent on the seed's energy reserves. The apical meristems start growing and give rise to the root and shoot. The first "true" leaves expand and can often be distinguished from the round cotyledons through their species-dependent distinct shapes. While the plant is growing and developing additional leaves, the cotyledons eventually senesce and fall off.

Consumption of Seedlings

Seedlings are commonly eaten as a health food. These seedlings are usually labeled sprouts, but in a botanical sense are actually seedlings. There is controversy in whether seedlings, or sprouts, are really worth eating. These seedlings are used many times in organic foods and are eaten for their concentrations of certain vitamins the seedling naturally contains.

This can be viewed out of proportion as the seedling is usually too small to contain enough vitamins or minerals to be in line with the amount they are claimed to obtain.

SPROUTING

Sprouting is the practice of soaking, draining and then rinsing seeds at regular intervals until they germinate, or sprout. This can be a semi - automated or fully automated process when done on a large scale for commercial use.

Seeds that can be Sprouted

One of the most common sprouts is that of the mung bean (*Vigna radiata*), often sold as 'Chinese Bean Sprouts'; another common sprout is the alfalfa sprout and the barley sprout.

Other seeds that can be sprouted include adzuki bean, almond, amaranth, annatto seed, anise seed, arugula, basil, navy bean, pinto bean, lima bean, broccoli, buckwheat, cabbage, canola seed, caragana, cauliflower, celery, chia seed, chickpeas, chives, cilantro (coriander, dhania), clover, cress, dill, fennel, fenugreek, flax seed, garlic, hemp seed, kale, kamut, kat, leek, green lentils, lupins,pearl millet, mizuna, mustard, oats, onion, black-eyed peas, green peas, pigeon peas, snow peas, peanut, psyllium, pumpkin, quinoa, radish, rye, sesame, soybean, spelt, sunflower, tatsoi, triticale, watercress, and wheat berries.

However, many sprouts are in fact toxic when eaten, like kidney beans. Some sprouts can be cooked to remove the toxin, while others will be toxic either way and should be avoided. So before eating any sprouts, find out if that species is edible as a sprout.

With all seeds, care should be taken that they are intended for sprouting or human consumption rather than sowing. Seeds intended for sowing may be treated with chemical dressings. Several countries, such as New Zealand, also require that some varieties of edible seed be heat-treated, thus making them impossible to sprout.

Many varieties of nuts, such as almonds and peanuts, can also be started in their growth cycle by soaking and sprouting, although because the sprouts are generally still very tiny when eaten, they are usually called "soaks."

Moisture, warmth, and in most cases, indirect sunlight are necessary for sprouting. Some sprouts, such as mung beans, can be grown in the dark. Little time, effort or space is needed to make sprouts.

To sprout seeds, the seeds are moistened, then left at room temperature (between 13 and 21 degrees Celsius) in a sprouting vessel. Many different types of vessels can be used. One type is a simple glass jar with a piece of cloth secured over its rim. 'Tiered' clear plastic sprouters are commercially available, allowing a number of "crops" to be grown simultaneously. By staggering sowings, a constant supply of young sprouts can be ensured. Any vessel used for sprouting must allow water to drain from it, because sprouts that sit in water will rot quickly. The seeds will swell and begin germinating within a day or two.

Sprouts are rinsed as little as twice a day, but possibly three or four times a day in hotter climates, to prevent them from souring. Each seed has its own ideal sprouting time. Depending on which seed is used, after three to five days they will have grown to two or three inches in length and will be suitable for consumption.

If left longer they will begin to develop leaves, and are then known as baby greens. A popular baby green is sunflower after 7-10 days. The growth process of any sprout can be slowed or halted by refrigerating until needed.

Common causes for sprouts to turn out inedible

- Seeds are allowed to dry out
- Seeds are left in standing water
- Temperature is high or too low
- Insufficient rinsing
- Dirty equipment
- Insufficient air flow
- Contaminated source of water
- Poor rate of germination of seed

These problems are easily solved by an automatic sprouter that mists and drains the sprouts at regular intervals. To control temperature, in the winter a warming blanket can be placed under the sprouter, and in the summer small fans in the lid if it's very hot and humid.

Mung beans can be sprouted either in light or dark conditions. Those sprouted in the dark will be crisper in texture and whiter, as in the case of commercially available Chinese Bean Sprouts, but these have less nutritional content than those grown in partial sunlight.

Growing in full sunlight is not recommended, because it can cause the beans to overheat or dry out. Subjecting the sprouts to pressure, by placing a weight on top of them in their sprouting container, will result in larger, crunchier sprouts similar to those sold in Polish grocery stores.

A very effective way to sprout beans like lentils or adzuki is in colanders. Soak the beans in hot water for about 8 hours then place in the colander. Wash twice a day. The sprouted beans can be eaten raw or cooked.

NUTRITIONAL INFORMATION

Sprouts are rich in digestible energy, bioavailable vitamins, minerals, amino acids, proteins, beneficial enzymes and phytochemicals, as these are necessary for a germinating plant to grow. These nutrients are essential for human health. To clarify the nutritional changes upon germination & sprouting have been summarized below.

Chavan and Kadam concluded that - "The desirable nutritional changes that occur during sprouting are mainly due to the breakdown of complex compounds into a more simple form, transformation into essential constituents, and breakdown of nutritionally undesirable constituents."

"The metabolic activity of resting seeds increases as soon as they are hydrated during soaking. Complex biochemical changes occur during hydration and subsequent sprouting. The reserve chemical constituents, such as protein, starch and lipids, are broken down by enzymes into simple compounds that are used to make new compounds."

"Sprouting grains causes increased activities of hydrolytic enzymes, improvements in the contents of total proteins, fat, certain essential amino acids, total sugars, B-group vitamins, and a decrease in dry matter, starch and anti-nutrients. The increased contents of protein, fat, fibre and total ash are only apparent and attributable to the disappearance of starch.

However, improvements in amino acid composition, B-group vitamins, sugars, protein and starch digestibilities, and decrease in phytates and protease inhibitors are the metabolic effects of the sprouting process."

INCREASES IN PLANT ENZYME CONTENT

Sprouts are a tremendous source of (plant) digestive enzymes. Enzymes act as biological catalysts needed for the complete digestion of protein, carbohydrates & fats. The physiology of vitamins, minerals and trace elements is also dependant on enzyme activity.

"Being eaten whilst extremely young, "alive" and rapidly developing, sprouts have been acclaimed as the "most enzyme-rich food on the planet". Estimates suggest there can be up to 100 times more enzymes in sprouts than in fruit and vegetables, depending on the particular type of enzyme and the variety of seed being sprouted. The period of greatest enzyme activity in sprouts is generally between germination and 7 days of age."

"Grains and legume seeds of all plants contain abundant enzymes. However, while grains and seeds are dry, enzymes are largely inactive, due to enzyme inhibitors, until given moisture to activate germination. It is these inhibitors that enable many seeds to last for years in soil without deteriorating, whilst waiting for moisture.

Enzyme inhibitors in some grains and legume seeds need to be inactivated by heating or other processes, before they can be safely fed. However, heating, cooking and grinding processes can also inactivate certain digestive enzymes within grains and seeds. Fortunately, during germination and sprouting of grains and seeds, many enzyme inhibitors are effectively neutralized, whilst at the same time the activity of beneficial plant digestive enzymes is greatly enhanced."

Increases in Crude Protein content Morgan et al. found that - "The protein content of sprouts increased from the time of germination. The absorption of nitrates facilitates the metabolism of nitrogenous compounds from carbohydrate reserves, thus increasing crude protein levels."

Increases in Protein Quality Chavan and Kadam stated - "Very complex qualitative changes are reported to occur during soaking and sprouting of seeds. The conversion of storage proteins of cereal grains into albumins and globulins during sprouting may improve the quality of cereal proteins. Many studies have shown an increase in the content of the amino acid Lysine with sprouting."

"An increase in proteolytic activity during sprouting is desirable for nutritional improvement of cereals because it leads to hydrolysis of prolamins and the liberated amino acids such as glutamic and proline are converted to limiting amino acids such as lysine."

Increases in Crude Fibre content Cuddeford, based on data obtained by Peer and Leeson, stated - "In sprouted barley, crude fibre, a major constituent of cell walls, increases both in percentage and real terms, with the synthesis of structural carbohydrates, such as cellulose and hemicellulose".

INCREASES IN ESSENTIAL FATTY ACIDS

An increase in lipase activity has been reported in barley by MacLeod and White, as cited by Chavan and Kadam.

Increased lipolytic activity during germination and sprouting causes hydrolysis of triacylglycerols to glycerol and constituent fatty acids.

Sprouting treatment of cereal grains generally improves their vitamin value, especially the B-group vitamins. Certain vitamins such as á-tocopherol (Vitamin-E) and â-carotene (Vitamin-A precursor) are produced during the growth process.

Sprouts provide a good supply of Vitamins A, E & C plus B complex. Like enzymes, vitamins serve as bioactive catalysts to assist in the digestion and metabolism of feeds and the release of energy. They are also essential for the healing and repair of cells. However, vitamins are very perishable, and in general, the fresher the feeds eaten, the higher the vitamin content.

The vitamin content of some seeds can increase by up to 20 times their original value within several days of sprouting. Mung Bean sprouts have B vitamin increases, compared to the dry seeds, of - B1 up 285%, B2 up 515%, B3 up 256%. Even soaking seeds overnight in water yields greatly increased amounts of B vitamins, as well as Vitamin C. Compared with mature plants, sprouts can yield vitamin contents 30 times higher."

Chelation of Minerals Shipard claims that - "When seeds are sprouted, minerals chelate or merge with protein, in a way that increases their function."

Risks and Antinutritional factors Some legumes can contain toxins or antinutritional factors, which can be reduced by soaking, sprouting and cooking (eg, stir frying). Joy Larkcom advises that to be on the safe side "one shouldn't eat large quantities of raw legume sprouts on a regular basis, no more than about 550g daily". However not all legume sprouts contain these antinutritional factors and many have greatly beneficial properties so it is recommended that the advice of a qualified nutritionist is sought before making any decisions about what to include or eliminate from a diet/ration.

Reduction of Anti-Nutritional Factors Phytic Acid occurs primarily in the seed coats and germ of plant seeds. It forms insoluble or nearly insoluble compounds with minerals including Calcium, Iron, Magnesium and Zinc, such that they cannot be effectively absorbed into the blood. Diets high in phytic acid and poor in these minerals produce mineral deficiency symptoms in experimental animals. The latter authors state that the sprouting of cereals has been reported to decrease levels of Phytic Acid. Similarly, Shipard states that enzymes of germination and sprouting have the ability to eliminate detrimental substances such as Phytic Acid.

Buckwheat greens contain fagopyrin, a naturally occurring substance in the buckwheat plant. When ingested in sufficient quantity, fagopyrin is known to cause the skin of animals and people to become phototoxic, which is to say hypersensitive to sunlight, particularly if juiced or eaten in large quantities. Due to the growing popularity of sprouts in general, and a widespread ignorance as to the toxic dangers posed by buckwheat greens specifically, many people are today suffering unnecessarily.

SPROUTING AND THE LIVING FOODS DIET

Advocates of a raw food diet promote the use of sprouting as an effective way to increase the nutrient value, and digestibility, of beans, seeds and nuts. Sprouts are in fact the most nutrient dense food on earth– nutrient density refers to the amount of nutrients per calorie. This is why many people on a low-calorie diet make sprouts and baby greens the mainstay of their diet, especially blended for easier digestion.

Many raw food dietitians avoid grains in favour of nuts and seeds. While raw nuts and seeds contain enzyme deactivators that harm the stomach and cooking nuts and seeds denatures their fats and oils, soaked or sprouted nuts and seeds have natural oils and active enzymes at the same time.

One main criticism of the raw foods diet, especially the insistence on the health value of sprouted nuts, seeds, and occasionally grains, is the very intensive labour time require for food preparation. Not only does it take a long time and lots of labour to sprout nuts and seeds, but the sprouting process makes them not as shelf-stable.

Furthermore, unlike most cooked foods, there are very few commercial avenues for purchasing such foods. Most "raw foods bars" are raw but not sprouted (and therefore do not have active enzymes) because the bars would not keep as long on the shelf. The same is true of most nut and seed butters.

A seed is essentially a baby in a suitcase carrying its lunch. "Baby" refers to the embryo, or immature plant, that will grow and develop into the seedling and ultimately the mature plant. The "suitcase" is the seed coat (or testa) that surrounds the seeds and "lunch" refers to the nutritive source for the germinating seedling. The food for the germinating seedling may be stored in part of the embryo itself, such as the fleshy cotyledons of a bean seed, or it may take other forms including endosperm, which is a special starch-rich storage tissue that surrounds the embryo.

A seed is officially considered to have germinated when the young root, called the radicle, emerges from the seed coat. To germinate, a seed requires three things – water, oxygen, and a suitable temperature. Water uptake, also called imbibition, is the first stage of seed germination. During this process the dry seed, which typically has a water content of less than 10%, absorbs water and swells.

This process serves to hydrate the dry components of the seed and active the metabolic machinery necessary for germination. Among the early metabolic activities occurring in the seed is the breakdown of starches stored in the seed into simple sugars that can be used for energy and building blocks for necessary cellular structures. Except for the first half-hour or so of germination when little oxygen is present, seed germination and subsequent seedling growth requires oxygen. It is required, in large part, for use in the cellular structures, called mitochondria, to produce ATP (energy).

A suitable temperature is necessary to optimize the metabolic reactions required for germination. The seeds of every species have an optimal temperature for germination; some species, such as the gourds and squashes prefer warm temperatures while other species such as radish can tolerate cooler temperatures for germination.

A seed that has not germinated because it is lacking one or more of the necessary requirements for germination is termed quiescent. These seeds are simply "resting", waiting for the appropriate conditions for germination. Given water, oxygen and/or a suitable temperature, a quiescent seed will germinate. However, even if given the proper conditions, a seed may not germinate. These seeds may fail to germinate because the seed is either dormant or "dead".

Dormant seeds have the potential to germinate but are prevented from doing so by some mechanism. Thus, even though all the proper growth conditions are present, they don't germinate unless they have been "primed" and there dormancy mechanism has been overcome. There are many dormancy mechanisms in seeds.

When some seeds, like hemp, are shed they have immature embryos that will not germinate until they undergo a period of development (called after-ripening). Other seeds, like apple, require a cold treatment, called stratification, for germination. Many of our native plant seeds must be stratified. Some seeds have a hard seed coat that needs to be nicked (called scarification) for germination. This usually occurs as the result of natural freeze-thaw cycles. Still other seeds require a period of heat in order to germinate. Many of these species are winter annuals that germinate in the late summer/early fall. Ultimately, the function of these varied dormancy mechanisms is to enable the seed time to disperse from the parent plant and to avoid germinating during unfavourable weather.

Humans have attempted to breed dormancy mechanisms from our crop plants. Although an advantage for a wild plant, dormancy is a problem if a farmer who want the crop to germinate uniformly and immediately upon planting. It's not easy to tell if a seed is "dead". Only if it fails to germinate when provided the proper conditions and any dormancy mechanisms are broken can we consider a seed "dead".

SEED GERMINATION

Plants produce many fruits and therefore many seeds.

This is because only a few of these seeds will find suitable places and conditions in which to germinate. The majority of the seeds will be wasted; some wind-dispersed seeds could fall in water or on stones.

When fruits and seeds are dispersed the seeds have very little water in them. The outer covering, the testa, is hard and sometimes rough and creased. Although the seeds seem dry and inactive they are very much alive. They are said to be in a dormant state and some seeds can remain like this for a long period of time. Inside the testa there is just enough water to keep the embryo (the baby plant) living until the external conditions are correct for the seed to germinate.

The Structure of a Haricot Bean Seed

The outer covering of a seed is called the testa. It is usually hard and protects the softer parts of the seed. There is a tiny hole in the testa called the micropyle. When the seed is ready to germinate, water is taken in through the micropyle. The first root, called the radicle, will grow out of the seed through the micropyle. On the surface of the testa there is a scar which shows the place where the seed was attached to the fruit.

A dried seed is difficult to study. If the seed is soaked in water for twelve hours it takes in some of the water and swells. Peas and beans are useful seeds to study after they have been soaked because they are big. The testa can be removed easily and the inside of the seed can be examined under the binocular microscope.

INSIDE THE SEED

Inside both the pea and the bean there are two large food-storing structures called cotyledons. The food contained in the cotyledons gives energy to the embryo when the seed germinates. The embryo is found between the two cotyledons. It is made up of two parts; the radicle, or first root, which points toward the micropyle, and the plumule, or first shoot. Under the binocular microscope the plumule can be seen as two very tiny leaves.

Some types of seeds have only one small cotyledon. Most

cereal grains, such as wheat, are like this. The embryo is surrounded by a structure called the endosperm which is also a food storage area. Both the cotyledon and the endosperm will provide energy to the embryo during germination.

SEED GERMINATION

A seed certainly looks dead. It does not seem to move, to grow, nor do anything. In fact, even with biochemical tests for the metabolic processes we associate with life (respiration, etc.) the rate of these processes is so slow that it would be difficult to determine whether there really was anything alive in a seed. Indeed if a seed is not allowed to germinate (sprout) within some certain length of time, the embryo inside will die. Each species of seed has a certain length of viability. Some maple species have seeds that need to sprout within two weeks of being dispersed, or they die. Some seeds of Lotus plants are known to be up to 2000 years old and still can be germinated.

Assuming the seed is still viable, the embryo inside the seed coat needs something to get its metabolism actived to start the embryo growing. The process of getting a seed to germinate can be simple or complicated, and this our present subject.

Seeds Lacking True Dormancy

Common vegetable garden seeds generally lack any kind of dormancy. The seeds are ready to sprout. All they need is some moisture to get their biochemistry activated, and temperature warm enough to allow the chemistry of life to proceed. Seeds taken from the wild, however, are frequently endowed with deeper forms of dormancy.

Thick Seed Coat

Many kinds of seeds have very thick seed coats. These obviously keep water out of the seed, so the embryo cannot get the water needed to activate its metabolism and start growing. The lotus seeds are an example of this. An outstanding example from the northern temperate zone is the

Kentucky coffee tree (*Gymnocladus dioica*). The seed coat is perhaps two millimeters thick! The Kentucky coffee tree holds its seed pods in the the top of the tree all winter. The inside of the pod is fleshy (lots of water). The pods are very dark in colour. Other species might use some pounding along a river or drop seeds into seacoast surf to abrade the thick seed coat. Some of the sea beans do this. Other seeds might need an vertebrate or other animal to attack the seed coat (but give up trying to eat the seed) and thereby weaken the coat. The process of nicking the thick seed coat to initiate germination is called scarification.

A final, and very common, example of a way to scarify a seed coat is observed in strawberry and raspberry. The thick seed coat is designed to be swallowed by the frugivore. The animal digests the fruit pulp, but the seed coat passes through the digestive system still protecting the viable embryo inside, but weakened enough to allow sprouting! The seed is deposited with a little organic fertilizer in the environment and can now sprout!

Thin Seed Coat

A thin seed coat is so thin that it is no barrier to water. Some other kind of dormancy mechanism is needed. Knowing that light can penetrate thin layers of plant tissue should give us the idea that light might be a signal. That plants can absorb light and respond biochemically is a fact we know from our study of photosynthesis. All we need is a pigment molecule that can absorb light and cause a change in the behaviour of the embryo.

The pigment is phytochrome. Like chlorophyll, it is made of a chromophore with tetrapyrole structure and is associated with proteins. This pigment is different from chlorophyll, however, in one critical way. It exists in two inter-convertible forms.

One form of phytochrome, named Pfr, is the form of the phytochome found in plant cells that are exposed to red (660 nm) or common white light. This form of phytochrome is biologically very active and plays a role in all systems when a plant needs to know if the lights are "on" or "off." In lettuce (*Lactuca sativa*) seeds, Pfr causes the seeds to begin to germinate as we will soon see. Thus lettuce seeds germinate only when placed in white or red light. Buried in deep soil, they will not germinate.

The other form of phytochrome, named Pr, is formed when phytochrome is exposed to far-red (730 nm) light. This form is biologically inactive or inhibits responses. Thus if lettuce seeds are placed in far-red light they do not germinate.

Large seeds have lots of storage material. If their seed coat is very thin, their evolution may have arrived at a completely different response. Think about pea seeds. They are large and have very thin seed coats. How would they respond to light?

Insufficient Development

If a seed's embryo is not completely developed, some additional maturation may be needed before the seed can sprout. This happens in seeds with little-to-no storage material invested in the seed. Examples include orchid seeds. They are the size of dust and have almost nothing but a very immature embryo on-board.

Such a seed needs an association with fungi in the soil or other environments to feed the developing embryo until the embryo is mature enough to actually penetrate the seed coat. These seeds are also likely to have a very brief viability. The fungal association must be established rapidly or the embryo dies.

Inhibitors Present

Many plant species invest chemicals in the developing seeds, and these chemicals inhibit the development of the embryos. They keep the embryos dormant. Obviously the seed must have some way to eliminate these chemicals before they can sprout.

Abscisic Acid

Many temperate zone species that use inhibitors use abscisic acid. This chemical induces dormancy in the embryo.

The chemical is produced in abundance in the late summer and early fall. The seeds in the fruits become dormant so, even if they are dispersed in autumn, they cannot sprout. During the winter enzymes in the seeds degrade the abscisic acid. By spring the abscisic acid is gone and the seed can sprout.

We can collect seeds of these species and get them to sprout early. The seeds are put in moist soil and refrigerated for about four weeks (a process often called stratification). This is sufficient time to degrade the abscisic acid. Then the planted seeds are placed in a warm greenhouse.

The seeds assume winter is over, spring has come, and they begin to sprout. This process is called vernalization. If we think of "vernal" as meaning "spring" then we understand how we got this name!

PHENOLIC COMPOUNDS

Plants that live in deserts have a different problem. There is no cold, moist, winter to allow vernalization of abscisic acid. These plants instead use more potent toxins, phenolic compounds, to keep their seeds dormant until the proper season for germination. Phenolic compounds are freely water-soluble, the plant is living in a desert.

Deserts typically have very long dry seasons and a short wet season accompanied by flash floods and so on. How do we think the phenolic compounds are lost? How would the mechanism ensure that seeds do not sprout in the dry season, but only after the seed could be sure it is in the wet season? The word leaching might give we a hint?

As any "typical" seed it has three fundamental parts

- A seed coat
- A storage area (in this case the endosperm) and
- A dormant embryo

The seed coat is really a fruit coat. In all grains, the seed coat (former ovule integument) is fused to the ovary wall (the true fruit wall). So in fact, the grain is technically a fruit (caryopsis) even though we often call it a seed.

Barley is used primarily in beer making. Brewers use barley as a source of sugar to make the alcohol for beer.

Brewers knew that sprouting the barley seeds improves the sugar yields tremendously, but they wanted even more sugar yield. The brewers came to plant physiologists to find out if anything could be done.

Study of the seed showed that the seed/fruit coat is water-resistant and thereby reduces the rate of water uptake by the seed, and water uptake is essential for seed germination and the improved sugar yields.

The sugar content of a dry barley seed is actually quite low, but the endosperm holds a huge reserve of starch. Starch is a polymer of sugars. In a sense it is a long chain of sugar molecules linked together. This is probably the source of the sugar.

The barley embryo has three parts

- The cotyledon (or seed leaf). Since there is only one, barley is a monocot.
- The epicotyl (becomes the shoot)
- The radicle (becomes the root)

Seed germination is said to have occurred when growth of the radicle bursts the seed coat and protrudes as a young root. The energy for seed germination probably comes from respiration of the sugar in the endosperm. However, the embryo and the starch are separated from each other. There must be some chemical communications that physiologists could manipulate to enhance sugar yield for the brewers.

The endosperm of the seed has two parts. The bulk of the volume is a starch storage area. The covering layer is called the aleurone layer. The aleurone is made of cells that store protein in abundance.

We probably know about this in the case of rice. Rice grains are treated before we buy them for cooking. The fruit/seed coat is milled off and discarded as it has little food value. The embryo breaks out and is called the "germ." In wheat, the embryos are sold as wheat germ. The milling process leaves the rice endosperm surrounded by aleurone. This is sold as brown rice and has food value in starch and protein. It has an interesting texture after cooking. If the endosperm is polished before it is sold, the aleurone can be removed. Polishing results in white rice which has food value in starch only. Thus white rice, while considered much "finer" than brown rice, is actually a lower-quality food.

The first step in barley seed germination is imbibition. In this process, water penetrates the seed coat and begins to soften the hard, dry tissues inside. The water uptake causes the grain to swell up. The seed/fruit coat usually splits open allowing water to enter even faster. The water begins to activate the biochemistry of the dormant embryo.

The water coming into the seed and embryo dissolves a chemical made inside the embryo. This chemical is called Gibberellic Acid (GA). It is a plant hormone, not too different from steroids. The dissolved GA is transported with the water through the rest of the seed tissues until it arrives at the aleurone layer. The GA crosses into the cytoplasm of the aleurone cells and turns on certain genes in the nuclear DNA. DNA is, of course, the hereditary molecule and contains the instructions for making every protein needed for the survival of a barley plant.

The precise mechanism of how GA turns on the DNA is unknown at present. It is clear however that the mode of action is to turn on just *certain* genes in the DNA. The genes that are turned on are transcribed. The information archived in the DNA is precious, so the aleurone cells make a disposable RNA copy of the gene that is turned on. This disposable copy of the information, a kind of blueprint, is often called messenger RNA. The process of making this RNA copy is called transcription. The RNA that was made in the transcription process is transported into the cytoplasm of the aleurone cells.

In the cytoplasm, the messenger RNA joins up with a ribosome to begin the process of making a protein. This process is often called protein synthesis or translation. In this process, the ribosome examines the information held in the sequence of bases in the RNA. Transfer RNAs charged with particular amino acids are moved into the position specified by the instructions in the messenger RNA, and the amino acids are joined in a proper sequence by the ribosome. The sequence of amino acids determines the properties of the protein being assembled.

In this case, the critical protein made with the information held in the RNA is amylase. This protein turns out to be an enzyme of great importance. The process of information held in the genes of DNA being transcribed into RNA and then translated into protein constitutes the central dogma of genetics.

A question one might ask right away is from where do the amino acid building blocks for the amylase come?

The answer is from some other biochemistry in the aleurone cells. This biochemistry causes the storage proteins in the aleurone cells to be digested by hydrolytic enzymes. The hydrolysis is accelerated by enzymes known as proteases. These enzymes increase the rate at which the storage protein is cut into individual amino acids.

The amino acids released by the hydrolysis are then free to be reassembled by the ribosomes into the structure of amylase. The same thing happens in people. In a similar way, barley aleurone storage proteins are digested and amylase is made from the released amino acids. Amylase is not just any old protein. It happens to be an enzyme. It catalyzes (accelerates) a particular chemical reaction. Specifically, the amylase speeds the hydrolysis of starch into its component sugar units. Hurrah! The physiologists now know how the sugar is made in a seed. They can also tell the brewers how to increase the sugar yields for beer making.

Note that Gibberellic Acid was the first plant hormone purified in large quantities to be successfully used to increase the yield of any economically-important biochemical in plants!

Remember that we wanted to get the seed germinated (sprouted). This does not occur until after the released sugar is transported from the endosperm to the embryo. The cotyledon is a major sugar transfer area in grains. Active sugar uptake is its specialized function. The sugar is transported into the embryo and used as a fuel and building block for growth of the embryo itself. The resulting embryo growth will include the emergence of the radicle from the seed/fruit coat.

Our barley seed story is great, but it does not explain some other forms of seed germination. The next great finding in the study of seed germination is what happens in lettuce seeds (*Lactuca sativa*). Lettuce seeds look somewhat different from barley seeds, because they are dicots. They also have separate seed coat and fruit walls. But the things we buy as lettuce "seeds" are really lettuce fruits (called cypselas). As different as they are, the seed germination process in lettuce needs just one step more than barley to get its seeds germinating.

Lettuce cypselas have very thin fruit walls and seed coats, so light can penetrate these coverings. As is typical of small seeds, light is needed to stimulate seed germination. One might wonder how this would fit into what we just found out with barley. The DNA must first be photoactivated before GA can have its effect. This is accomplished by the formation of a chemical signal known as Pfr. Pfr is one of two possible chemical forms of a molecule known as phytochrome. Phytochrome is a pigment that absorbs light energy. When it does absorb light energy its internal chemical structure is altered instantly.

Thus Pfr can be converted to the alternate Pr form of phytochrome by simply hitting the seeds with a flash of far-red light (730 nm). Similarly, the Pr form of phytochrome can be converted back into Pfr with a flash of red light (660 nm). Dry lettuce seeds have slightly more than half of their phytochrome in the Pfr form. So if they are moistened and kept in the dark, some of the seeds will actually germinate. If we shine red (or it turns out even white) light on imbibing seeds, then virtually all of the phytochrome is converted to the Pfr form and this photoactivates the genes in the DNA. Almost all of the seeds will germinate in this light!

On the other hand, if we shine far-red light on imbibing seeds, then virtually all of the phytochrome in the seeds is converted to the Pr form and none of the seeds will germinate. Thus phytochrome and light tell lettuce seeds when they are exposed to sunlight and can begin to germinate. Small seeds, such as those in lettuce, do not have reserves enough to start germinating without being able to quickly get to light. It is no surprise that small-seeded species have evolved this mechanism for photoactivation of seed germination. It prevents a deeply-buried seed from trying to germinate!

Chapter 5

Chemistry of Photosynthesis

The ultimate source of energy for life on Earth is the sun. Plants are able to transform the light energy from the sun into chemical energy through a process called photosynthesis. Through absorption of light energy, plants, algae, and a few types of bacteria transform carbon dioxide and water into carbohydrates. Carbohydrates serve as fuel for their growth and metabolism. Other organisms benefit indirectly from photosynthesis. Oxygen is a waste product of photosynthesis. By releasing oxygen into the environment, plants make the air breathable for many other life forms.

The presence of oxygen in relation to plants was first demonstrated by the English chemist, Joseph Priestley in 1772. He noted that air that had been depleted through burning candles could be made breathable again by plants. Seven years later, a Dutch doctor, Jan Ingen- Housz, showed that plants required sunlight in order to make air breathable. Ingen-Housz also demonstrated that it was only the green parts of plants that had this ability. However, neither scientist knew anything of oxygen's existence.

In 1782, Swiss scientist Jean Senebier proved that the plants also needed "fixed air"–what is now called carbon dioxide–in addition to sunlight. His discovery was followed in 1804 by another Swiss scientist, Nicholas Théodore de Saussure, proving that water was also a necessary factor. By the early 19th century, scientists had at least a general understanding of photosynthesis. They knew that green plants, if given carbon dioxide, water, and sunlight, would produce organic material and oxygen. Building on this foundation, tists began to tease apart the details of how the plants accomplished this pocess.

A big step forward was the discovery that chloroplasts were the site of oxygen generation in plant cells. Chloroplasts are cell organelles comparable to the mitochondria of animal cells. These organelles contain a cell's energy-generating mechanisms. Discovering the function of chloroplasts prompted scientists to scrutinize the organelle's structure and chemical makeup.

Not all cells capable of photosynthesis have chloroplasts, however. Cells are generally divided into two categories prokaryotes and eukaryotes. Eukaryotes have chloroplasts, but prokaryotes do not. The difference between eukaryotes and prokaryotes is that eukaryotes have a nucleus, while prokaryotes do not. The nucleus is a cellular structure that stores genetic material separately from the rest of the cell interior. Plants and certain types of algae are eukaryotes. Blue-green algae and photosynthetic bacteria are prokaryotes. The following discussion of photosynthesis will concentrate on photosynthetic eukaryotes–i.e., plants and algae.

Chloroplasts contain chlorophyll, a green-coloured pigment. Chlorophyll is the key molecule in photosynthesis. There are several types of chlorophyll, but the predominant form is chlorophyll *a*. Chlorophyll molecules are composed of a central porphyrin ring and side chains. These side chains differ slightly between types of chlorophyll, but the general appearance of the molecule is very similar. The porphyrin ring is actually a complex multi-ring structure composed of carbon, hydrogen, nitrogen, and oxygen atoms.

Light is broken down into units called photons which move in waves. The distance between waves is called the wavelength. Visible light contains wavelengths ranging from 350 nm to 800 nm. Longer wavelengths contain less energy; short wavelengths contain more. Wavelengths correspond with the colours of the rainbow–called the visible light spectrum–and each colour represents a different energy level.

The colours that an object appears to be are those that are reflected; colours that are absorbed are not seen. Chlorophyll absorbs energy from the red portion of the spectrum, approximately 680-700 nm, and reflects the green portion. For that reason, the leaves and stems of plants appear green. When a photon is absorbed by a chlorophyll molecule, its energy causes the chlorophyll to enter an energy-rich excited state.

Within a chloroplast, chlorophyll molecules are densely packed together. Because the chlorophyll molecules are so close to each other, the energy they absorb circulates from molecule to molecule. This energy is eventually transmitted to a reaction centre, thought to be a special arrangement of chlorophyll and protein molecules. There are two types of reaction centres in chloroplasts, each of which corresponds with a photosystem. The P700 reaction centre is associated with photosystem I; P680 is linked with photosystem II. The names of the photosystems arise from the wavelength of light which they most efficiently absorb.

Both photosystems serve as trigger points for electron transport chains. Electron transport chains operate through the reduction and oxidation of chemicals or chemical complexes that make up its links. Although reduction usually means that something is lost, it has the opposite meaning in chemistry. As a chemical term, reduction means that an electron is gained. Oxidation means that an electron is lost.

When P680 absorbs the energy of a photon, it goes from a ground state to an excited state. This change is seen at the atomic level. In the ground state, the electrons of an atom are at a low- energy configuration. In an excited state, an electron jumps from a normal low-energy position, or orbital, to a high-energy orbital. The high-energy position is unstable, and the electron is quickly transferred to another chemical, pheophytin. Pheophytin is the first link in the photosystem II electron transport chain.

As P680 returns to its ground state, it picks up an electron to replace the one transferred to pheophytin. This electron is taken from a water molecule. In chemical terms, P680 is reduced and water is oxidized. The water molecule is split into hydrogen ions (positively charged atoms) and oxygen. For every two water molecules broken down in this manner, four hydrogen ions and one oxygen molecule are produced. The hydrogen ions can be used for other reactions, but the molecular oxygen is released as a useless waste product.

The electron transport chain that picks up an electron from P680, leads to P700. The chain itself comprises several chemical links. The extra electron is passed from one link to the next in a series of reductions and oxidations. As the electron proceeds to P700, it encounters a chemical complex called cytochrome *bf*. The final link in the photosystem II electron transport chain is plastocyanin.

The P700 reacts to a photon in the same manner as P680. When the photon's energy is absorbed, P700 is transformed from its ground state to an excited state. In its excited state, P700 transfers an electron to the photosystem I electron transport chain. In returning to its ground state, P700 is reduced by plastocyanin, courtesy of photosystem II.

The photosystem I electron transport chain leads to an enzyme called ferredoxin-NADP reductase. (Enzymes are protein molecules that trigger speedy chemical reactions.) Once this enzyme has the extra electron available, it is able to transform NADP+ to NADPH and a hydrogen ion. NADPH is the abbreviation for nicotinamide adenine dinucleotide phosphate. This molecule is a high-energy fuel used for certain chemical reactions in the cell.

Under certain circumstances, P700 is raised to its excited state and the electron passes through only the first few links of the photosystem I electron transport chain. Instead of proceeding to the last link, the electron is shunted over to the cytochrome *bf* complex of photosystem II. The end result of this process–called cyclic photophosphorylation–is the production of a different high-energy fuel. This fuel is adenosine triphosphate, or ATP. Both ATP and NADPH are use to drive the chemical reactions that produce carbohydrates.

Both photosystem I and photosystem II require sunlight. As a result, this part of photosynthesis falls under the heading of light reactions. The actual generation of carbohydrates is a dark reaction, because it can occur in the absence of light. Carbohydrate construction, or synthesis, is accomplished through the Calvin cycle. The Calvin cycle is a series of chemical reactions through which carbon dioxide is used to build glucose. Glucose, in turn, is used as an eventual building block for sucrose, starch, and other carbohydrates.

The first step of the Calvin cycle is the addition of carbon dioxide to an acceptor molecule. This step was first described by Melvin Calvin in the 1940s, and the entire cycle is named in his honor. The cycle begins with the addition of carbon dioxide to another molecule, ribulose-1,5-bisphosphate. This reaction is triggered, or catalyzed, the enzyme ribulose 1,5- bisphosphate carboxylase/oxygenase–often called by its short name, rubisco. Rubisco is able to add either carbon dioxide or molecular oxygen to a ribulose-1,5-bisphosphate molecule. The addition of carbon dioxide leads into the Calvin cycle; the addition of oxygen leads to a different process called photorespiration.

When rubisco binds carbon dioxide to ribulose-1,5-bisphosphate, a six-carbon molecule is temporarily formed. This molecule quickly splits into two molecules of 3-phosphoglycerate. At this stage, ATP enters the cycle by transferring a phosphate group to the molecule. The resulting molecule is reduced by NADPH to form glyceraldehyde-3-phosphate. At this point, glyceraldehyde-3-phosphate can go in one of two directions. A certain quantity of the molecule is dedicated to completing the cycle by regenerating the ribulose-1,5-bisphosphate stores. The remaining glyceraldehyde-3-phosphate continues on toward glucose, sucrose, starch, and other carbohydrate production.

Some plants have the ability to concentrate carbon dioxide levels within their cells before it enters the Calvin cycle. This ability helps circumvent photorespiration. Photorespiration occurs when rubisco adds oxygen, rather than carbon dioxide, to ribulose-1,5- bisphosphate. This is a wasteful reaction for the plant because it uses up ATP and reduces carbohydrate synthesis. Generally rubisco will use carbon dioxide in preference to oxygen, but in hot, sunny climates, the carbon dioxide near plants is rapidly used up. With lower levels of carbon dioxide in the area, oxygen uptake would increase and photorespiration would occur.

Some plants have evolved a mechanism to reduce photorespiration. These plants are called C4 plants, because they create a four-carbon molecule as an intermediary between carbon dioxide and the start of the Calvin cycle. The C4 plants concentrate carbon dioxide levels prior to the Calvin cycle through the Hatch-Slack pathway. The Hatch-Slack pathway is named for its discovers, Australian biochemists, Marshall Hatch and Roger Slack. Examples of C4 plants are corn, sorghum, and sugar cane, both of which do well in hot, sunny conditions.

Plants which add carbon dioxide directly to the Calvin cycle are called C3 plants, because they create the three- carbon molecule, 3-phosphoglycerate. The C3 plants do best in temperate climates. Wheat is an example of a C3 plant. Regardless of how carbon dioxide enters the Calvin cycle, the products are the same.

Photosynthesis evolved over three billion years ago, shortly after the appearance of the first living organisms. The food we eat and the oxygen we breathe are both formed by plants (including algae) through photosynthesis. The power to drive this reaction comes from sunlight absorbed by chlorophyll in the chloroplasts of plants. At the present time, no known chemical system can be made to serve as a substitute for this process.

It has been calculated that each CO_2 molecule in the atmosphere is incorporated into a plant structure every 200 years and that all the O_2 in air is renewed by plants every 2000 years. All life depends directly or indirectly on the sun's energy, and only plants are capable of capturing and converting this energy into chemical energy in the form of sugar and other organic compounds. Thus, if plants should suddenly disappear from the earth, so would we.

Our geological heritage of coal, oil, and gas also originated directly or indirectly from photosynthesis, since these fossil fuels were all derived from the remains of living organisms. Our stake in photosynthesis is, therefore, great, since we are not only dependent upon it for the food we eat, but also for many of the goods and most of the energy we use.

THE SITE OF PHOTOSYNTHESIS IN VASCULAR PLANTS

Leaves are the major organs of photosynthesis in vascular plants. Chloroplasts are found mainly in the mesophyll cells, the green tissue in the interior of the leaf

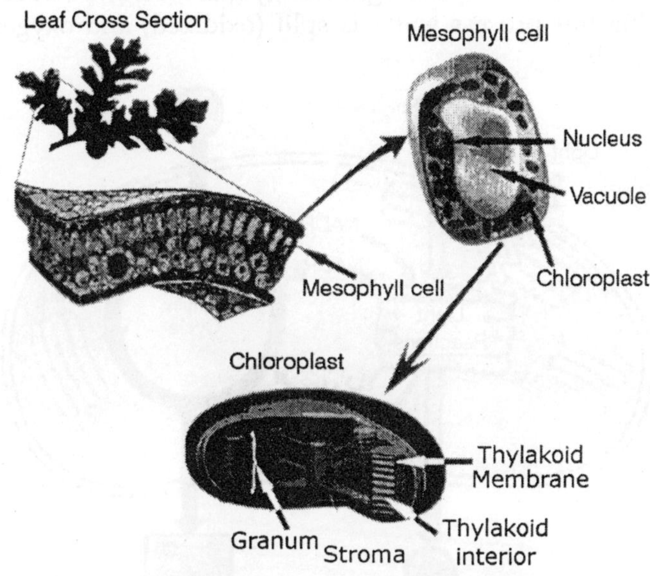

Fig Diagram of a Leaf Cross Section and Structure of the Chloroplast

Chloroplasts are ellipsoidal or disc-like in shape and are between 5 - 7 micrometers in diameter and 1 - 2 micrometers thick (1 micrometer = one millionth of a meter, 1 meter = 39 inches). There are about 36 chloroplasts in each mesophyll cell. The chloroplast is bounded by a double membrane that encloses the stroma, a dense aqueous solution that contains DNA, RNA, metabolites, and the enzymes associated with the conversion of CO_2 into organic matter. Membranes of the thylakoid system separate the stroma from the thylakoids. Thylakoids are concentrated in stacks called grana. Thylakoids contain the pigments chlorophyll *a* and *b*, carotenoid, and the enzymes associated with the oxidation (splitting) of water (H_2O) and the production of oxygen.

THE CHEMISTRY OF PHOTOSYNTHESIS

Photosynthesis takes place in the chloroplasts of plant cells and consists of Light-Dependent and Light-Independent reactions. The Light-Dependent reaction occurs in the thylakoids and converts light energy into ATP and $NADPH_2$. During this process water is split (oxidized) and oxygen is given off.

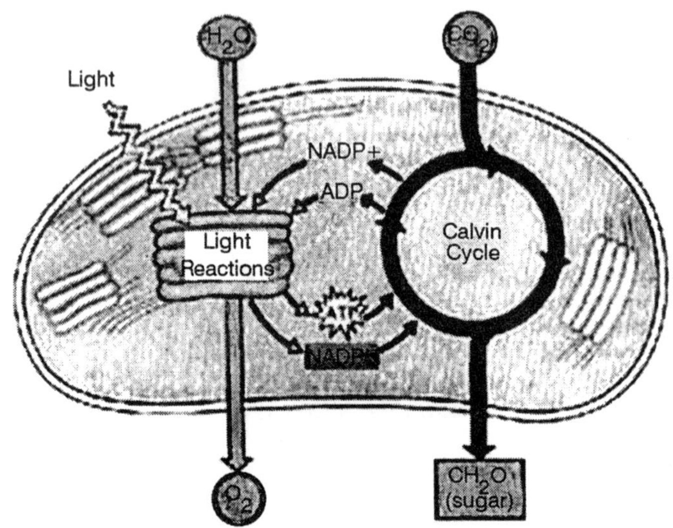

Fig Overview of Photosynthesis.

Light-Independent reactions (the Calvin Cycle) incorporate CO into sugar, the basic food source for all organisms. Thylakoid membranes are the sites of the Light-Dependent reactions, whereas the Calvin cycle occurs in the stroma. These reactions can be summarized by the following equations

Light-Dependent Reactions

$$\text{Thylakoids: } 2H_2O + 2NADP + 3ADP + 3P_1 \xrightarrow{\text{Light}} \underset{\text{Chlorophy11}}{\longrightarrow} O_2 + 2NADPH_2 + 3ATP$$

Light-Independent Reactions (Calvin Cycle)

$$\text{Stroma: } CO_2 + 2NADPH_2 + 3ATP \longrightarrow \longrightarrow \text{Sugar} + 2NADP + 3ADP + 3P_1$$

A DESIGN FLAW IN PHOTOSYNTHESIS – PHOTORESPIRATION

Since plants first moved onto land about 425 million years ago, they have been adapting to the problems of terrestrial life, particularly the problem of dehydration. The solutions often involve tradeoffs. An important example is the compromise between photosynthesis and the prevention of excessive water loss from the plant. The CO_2 needed for photosynthesis enters a leaf via microscopic pores called stomata.

However, the stomata are also the main avenues of transpiration, the evapouration of water from leaves. On hot, dry days, most plants close their stomata in order to conserve water. This response limits access to CO_2, thereby reducing photosynthetic yield. Under these conditions, CO_2 concentrations in the air spaces within the leaf begin to decrease and the concentration of oxygen released from photosynthesis begins to increase. This favours what appears to be a wasteful process within the leaf called photorespiration.

In most plants (about 800,000 species), CO_2 is initially fixed via rubisco of the Calvin cycle into ribulose bisphosphate. Because the product of this reaction is a three-carbon compound, 3-phosphoglycerate, such plants are called C_3 plants. Rice, wheat, and soybeans are among the C_3 plants that are important in agriculture. On hot, dry days when their stomata close, these plants produce less food as CO_2. levels within the leaf decline. Making matters worse, rubisco can use oxygen in place of CO_2. As O_2 increases, rubisco adds oxygen to RuBP instead of CO_2. The product splits, and one piece, a two-carbon compound (glycolate), is exported from the chloroplast. Other organelles (mitochondria and peroxisomes) then break down glycolate back to CO_2.

This process is called photorespiration because it occurs in the light (photo) and consumes oxygen (respiration). However, unlike normal cellular respiration, photorespiration generates no ATP. And unlike photosynthesis, photorespiration produces no food. Photorespiration is probably a metabolic relic from a much earlier time when the atmosphere had less O_2 and more CO_2 than it does today.

When rubisco first evolved, the inability of the enzyme's active site to exclude O_2 would have made little difference. Now, it is considered to be wasteful, since photorespiration drains away as much as 50% of the carbon fixed by the Calvin cycle. If photorespiration could be reduced in certain plant species, without affecting photosynthetic productivity, crop yields and food supplies would increase.

ALTERNATIVE STRATEGIES OF CARBON FIXATION - C_4 AND CAM C_4 PLANTS

The environmental conditions that promote photorespiration are hot, bright, dry days. In these climates, alternate modes of carbon fixation have evolved to minimize photorespiration. The two most important of these photosynthetic adaptations are exhibited by C_4 and CAM C_4 plants. C_4 plants are so named because they form a four-carbon compound as the first product of the nonlight requiring reactions of photosynthesis. Several thousand species in at least 19 families use the C_4 pathway. Agriculturally important C_4 plants are sugarcane and corn, members of the grass family.

Leaves of C_4 plants contain two distinct types of photosynthetic cells a cylinder of bundle-sheath cells surrounding the vein, and mesophyll cells located outside the bundle sheath. CO_2 is initially fixed in mesophyll cells by the enzyme PEP carboxylase. A four-carbon compound is formed (malate in this case) which conveys the fixed CO_2 via plasmodesmata (protoplasmic connections) into a bundle sheath cell where the enzymes of the Calvin cycle are located.

In the bundle sheath cell, the malate is converted into pyruvate and CO_2; the latter is now used by rubisco and the Calvin cycle to make sugar. Compared to rubisco, the enzyme PEP carboxylase has a much higher affinity for CO_2. Thus, PEP carboxylase can fix CO_2 efficiently when it is hot and dry and stomata are partially closed. Also, by pumping CO_2 from the mesophyll cells into the bundle sheath, this keeps the CO_2 concentration high enough for rubisco to accept CO_2 rather than O_2. In this way, C_4 photosynthesis minimizes photorespiration and enhances sugar production. This adaptation is especially advantageous in hot climates with intense sunlight and it is where C_4 plants evolved and thrive today.

A second photosynthetic adaptation to arid conditions (as found in deserts) has evolved in succulent (water-storing) plants (including ice plants), many cacti, and representatives of other plant families. These plants close their stomata in the day and open them during the night, just the reverse of other plants. Closing the stomata during the day helps desert plants conserve water, but it also prevents CO_2 from entering the leaves. At night, when the stomata are open, these plants take up CO_2 and initially fix it into four-carbon compounds like malate.

This mode of carbon fixation is called crassulacean acid metabolism, or CAM, after the plant family Crassulaceae, the succulents in which the process was first discovered. The photosynthetic cells of CAM plants store the malate formed in the night in their vacuoles until morning, when the stomata close. In the daytime, when the light reactions can make ATP and $NADPH_2$ for the Calvin cycle, CO_2 is released from the malate made the night before to become fixed into sugar in the chloroplasts.

The diagram compares C_4 and CAM C_4 photosynthesis. Both adaptations are characterized by initial fixation of CO_2 into an organic acid such as malate followed by transfer of the CO_2 to the Calvin cycle. In C_4 plants, such as sugarcane, these two steps are separated spatially; the two steps take place in two cell types. In CAM C_4 plants, such as pineapple, the two steps are separated temporally (time); carbon fixation into malate occurs at night, and the Calvin cycle functions during the day. Both C_4 and CAM C_4 are two evolutionary solutions to the problem of maintaining photosynthesis with stomata partially or completely closed on hot, dry days.

However, it should be noted, that in all plants, the Calvin cycle is used to make sugar from carbon dioxide. On a global scale, this represents a prodigious amount of sugar, about 160 billion metric tons of carbohydrate per year.

An example of naturally-occuring biological oxidation-reduction reactions is the process of photosynthesis. It is a very complex process carried out by green plants, blue-green algae, and certain bacteria. These organisms are able to harness the energy contained in sunlight, and via a series of oxidation-reduction reactions, produce oxygen and sugar, as well as other compounds which may be utilized for energy as well as the synthesis of other compounds. The overall equation for the photosynthetic process may be expressed as

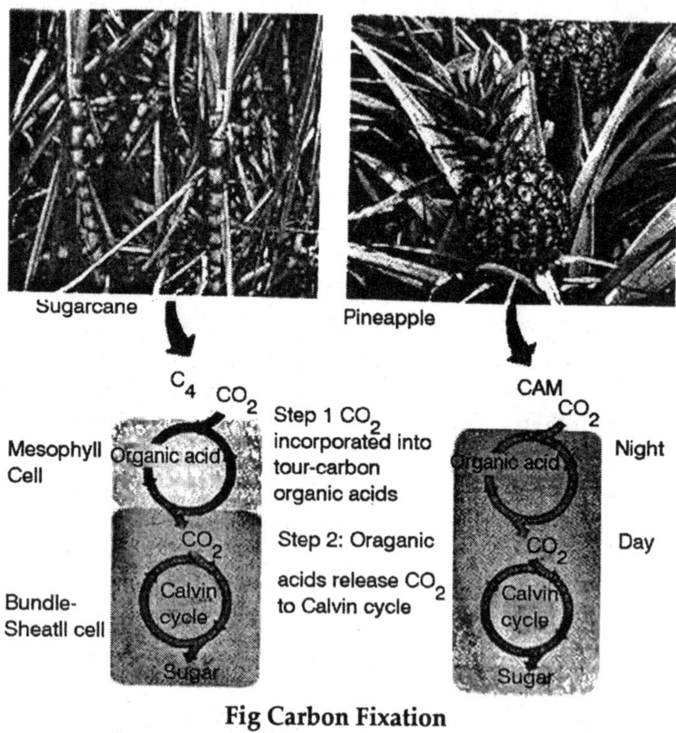

Fig Carbon Fixation

$$6\ CO_2 + 6H_2O \longrightarrow C_6H_{12}O_6 + 6\ O_2$$
(glucose)

The equation is the net result of two processes. One process involves the splitting of water. This process is really an oxidative process that requires light, and is often referred to as the "light reaction". This reaction may be written as

$$12\ H_2O \longrightarrow 6\ O_2 + 24\ H^+ + 24e^-$$
light or radiant energy

The oxidation of water is accompanied by a reduction reaction resulting in the formation of a compound, called nicotinamide adenine dinucleotide phosphate (NADPH). This reaction is illustrated below

$NADP^+ + H_20 \longrightarrow NADPH + H^+ + O$

(oxidized form) (reduced form) (oxygen)

This reaction is linked or coupled to yet another reaction resulting in the formation of a highly energetic compound, called adenosine triphosphate, (ATP). As this reaction involves the addition of a phosphate group (labeled, as P_i) to a compound called, adenosine diphosphate (ADP) during the light reaction, it is called photophosphorylation.

$ADP + P_i \longrightarrow ATP$

Think of the light reaction, as a process by which organisms "capture and store" radiant energy as they produce oxygen gas. This energy is stored in the form of chemical bonds of compounds such as NADPH and ATP. The energy contained in both NADPH and ATP is then used to reduce carbon dioxide to glucose, a type of sugar ($C_6H_{12}O_6$). This reaction, shown below, does not require light, and it is often referred to as the "dark reaction".

$6\ CO_2 + 24\ H^+ + 24\ e^- \longrightarrow C_6H_{12}O_6 + 6\ H_2O$

The chemical bonds present in glucose also contain a considerable amount of potential energy. This stored energy is released whenever glucose is catabolized (broken down) to drive cellular processes. The carbon skeleton in glucose also serves as a source of carbon for the synthesis of other important biochemical compounds such as, lipids, amino acids, and nucleic acids. In simplest terms, the process of photosynthessis can be viewed as one-half of the carbon cycle. In this half, energy from the sun is captured and transformed into nutrients which can be utilized by higher organisms in the food chain. The release of this energy during the metabolic re-conversion of glucose to water and carbon dioxide represents the second half of the carbon cycle and it may be referred to as catabolism or "oxidative processes"

HUMAN RESPIRATION

There are a number of books listed in the bibliography

that have excellent illustrations. Teachers might also consider use of a purchased or homemade chart of the process of respiration.

All human beings need air to survive. Actually, we need oxygen contained in the air. While food is also a necessity, we can survive for days without it. Without air, however, we would quickly die. During the process we call respiration we breathe in air into our lungs. When we release or exhale, carbon dioxide and water vapour are released. The oxygen we take in helps to oxidize or "burn food." Plants are the corner stone of this process. All living things depend upon food synthesized from raw materials in the leaves of green plants. Plants depend on their parts to help produce this food.

Humans and animals breathe through their lungs. As we breathe in air passes through the nasal passages above the mouth where it is warmed and filtered. The air then moves through the voice box or larynx where two tubes are found. One channels food into the stomach and the other called the windpipe goes into the lungs. When we eat or drink a flap of tissue covers the windpipe so we do not swallow food into our lungs.

From there the windpipe divides into two tubes called bronchi. At the end of each bronchus tube is one of our two lungs. Within the lungs are a series of progressively smaller branches of tubes. At the end of the tiniest tube is a tiny air sac. These sacs swell like tiny balloons when we breathe air into our lungs. Surrounding these sacs are many capillaries. Oxygen breathed into the sacs pass through the sacs' thin walls into the capillaries. Carbon dioxide and water vapour in the capillaries pass the opposite way into the air sacs. Children should remember oxygen in, carbon dioxide out.

THE EARTH'S ATMOSPHERE

Many children when talking about the air would simultaneously mention oxygen and air as the same thing. Our next discussion of the air should help to clear this up. At this point we have established that all humans and animals breathe. When we take in air what part or parts are we using and what are we exhaling? Is air the same as oxygen?

An ocean of air surrounds the earth. The air we breathe is a mixture of several different gases. The three most important that are needed for survival are oxygen, nitrogen, and carbon dioxide. We breathe in oxygen that combines with sugars in our body cells and releases heat energy. Students may believe that the air is pure oxygen. While we can breathe pure oxygen for short periods of time, too much pure oxygen is poisonous to our system and we would die.

Oxygen makes up about 21% of the air, and nitrogen 78%. There is really only a small amount of carbon dioxide in the air ñ about 3/100 of 1%. There are less than 1% of other gases such as argon, krypton, helium, neon, radon, and xenon. All the gases get mixed by the wind and are covering cover the earth in a five to six mile deep layer. Of course this mixture of gases is not all we breathe. There are other substances in the air like pollens, dust, smoke, salt particles, water vapour, chemicals, spores, bacteria and viruses.

Billions of years ago when the earth was first developing the air around the earth was made up of gases that evapourated from water that seeped out of the soil and escaped from the hot lava that came from beneath the earth's crust. The first atmosphere that the earth had would not have been able to sustain life, as we know it. When the first living organisms appeared in the water there was still no oxygen in the air. When the first plants appeared they changed the air by using up carbon dioxide and giving off oxygen.

PLANT ANATOMY

At this point, the unit has covered human respiration, air and its components. We have also gone over the basic gas exchange that happens when we breathe. We are now ready to turn to plants. My third grade students would invariably have studied plants in some way perhaps in preschool and/or in previous grades. As we turn the unit in the direction of plants, I would probably begin with a review of the basic parts of a plant and their functions. One way I might do this depending on my timeframe is to take my class to the library media centre to research plant parts. In that case, I might give them a copy of a plant drawing and a list of the parts and have them look them up, label the picture, and write the function of the parts.

We could also do this with tree parts and then compare and contrast parts of trees and parts of plants. I have not included in my list the parts of a flower but that can also be included depending on what the objectives are. As the unit moves on and we begin to talk about how plants breathe, the children might research the parts of a leaf. As a learning technique, this type of activity will usually result in the children retaining the subject matter longer and with better understanding than if they are given the diagrams already labeled.

BASIC PARTS OF A PLANT

Roots One of the basic functions of roots is to help anchor the plant into the ground. It is also through the roots that the plant absorbs water from the soil. The roots are also a storage place for food that the plant needs to store for those times when it cannot produce its own food. Stems These parts function like arteries carrying water from the roots throughout the plant. Minerals in the water are deposited in the leaves. When water reaches the leaves some of it evapourates into the air. Leaves Inside leaf cells are green-pigmented chemicals, chlorophyll. Chlorophyll enables a plant to combine carbon dioxide into a simple sugar. The energy to do this comes from sunlight that is absorbed by the chlorophyll. The process called photosynthesis literally means, "to put together with light."

From the sugar the plants make starch that they store throughout the plant. With compounds from the soil and soil bacteria, plants can manufacture vitamins and proteins. The size of plants leaves is a good indicator of how quickly water will evapourate from it. Broad leaf plants like maple will evapourate water quicker than other smaller leafed plants. Desert plants usually have thick, round leaves which reduce the surface and thereby the evapouration rate. They also have a waxy covering, which inhibits evapouration.

PHOTOSYNTHESIS

So far we have talked in general terms about the breathing process of animals and humans. Now the aim is to talk about the breathing process in plants. We now want to get more specific about what happens during the breathing process of plants. The teaching point here is for students to see the reciprocity between the two life forms. We will begin with animals and then spend more time on what the exchange of gases is like in plants; this should lead into a direct discussion of photcsynthesis.

How do Plants Breathe?

Essentially the leaves on a plant act much like lungs. Thousands of microscopic openings called stomata are located largely, but not exclusively on the underside of leaves. Each stomata is surrounded by special cells, which regulate the size of the opening. Water in the leaf evapourates into the air transpiration. The stomata play an important part in a plants' survival. In dry spells and at night the stomata can close and keep the plant from losing moisture. In the procedure, a few leaves on a plant are covered with Vaseline. After leaving the plant for a couple of days, the leaves will begin to wilt and die. Oxygen is not being released and carbon dioxide is not entering the problem so the leaves' system is backing up.

The word photosynthesis means, "putting together with light." Plants do this when they use carbon dioxide, water, and sunlight to produce the food they need to survive. Photosynthesis takes place mainly in a plant's leaves. Because leaves are so thin no plant cell is very far from the surface. When the sunlight hits the leaves the leaf absorbs the light. Cells in leaves contain chloroplasts that contain a green pigment called chlorophyll.

This material gives leaves their green colour. When light is absorbed the electrons in the chlorophyll become so filled with energy that they break out of their molecule. Every chlorophyll molecule that loses an electron is incomplete. The only way it can be complete is to take an electron from a water molecule. When an electron is taken away from a water molecule the molecule splits into hydrogen and oxygen. The oxygen returns back into the air where animals and humans breathe it. The hydrogen is used to make glucose a simple form of sugar. The released electron is used to make ATP (adenosine triphosphate). ATP is used by all living things to store, carry, and transfer energy. It has three phosphorus atoms bonded together. When energy is needed the last bond is broken and ATP becomes ADP. In this process, light energy becomes chemical energy that the cells in the plant can use. Carbon dioxide comes into the plant when the stomata are open. It provides the oxygen and hydrogen atoms needed to make glucose. The glucose is made through a series of reactions that are driven by the energy released from the ATP molecule. When the ATP molecules lose their energy they become ADP.

The ATP can be reenergized if the ADP can pick up more energy and reform its bonds. They return to be reenergized. Glucose is used by the plant to make cellulose and fats. The Glucose that is not used is stored as starch. This is the basic food used by plants. Most important to human beings and other life on earth is that in this process of photosynthesis, plants use less than they produce. That left over is returned to the atmosphere. Therefore the plants and trees around us help to balance the gases in our air by using up carbon dioxide and making oxygen. In the case of plants, the opposite of humans and animals is true - carbon dioxide in, oxygen out.

ECOSYSTEMS

Before we begin talking about the social issues involved with photosynthesis it would be helpful here to explain to students that the earth is considered a closed system. The earth is referred to as a closed eco system because all of the materials we need for life are here. We get no new infusions of material and so the earth must continually recycle these materials so that life can continue. We have already spoken about one of the cycles the carbon dioxide-oxygen cycle. It might be useful to also mention and illustrate two other important cycles.

NITROGEN CYCLE

Nitrogen is also an important element in a living cell.

There is nitrogen in the air but plants and humans cannot use it in the gaseous form. Plants get nitrogen when bacteria attach themselves to the roots of plants. The bacteria combine gaseous nitrogen and hydrogen to make a chemical compound called ammonia. Plants absorb the ammonia. Animals eat plants and get the nitrogen they rely on.

WATER CYCLE

Water on the earth's surface continually condenses in the sky and falls again as rain, snow or sleet. Living things need water to carry chemicals to cells and remove waste materials from cells. Plants release water through transpiration, a process in which water vapour is given off through leaves and into the air. Animals release water through exhalation, perspiration, and in waste products. This released water goes into the air and becomes part of the water cycle.

SAVING THE EARTH

Having now seen how the plant and animal life are tied in this reciprocal relationship it makes us aware of the necessity to take care of the earth for future generations and ourselves. As part of our third grade curriculum on communities, we would be studying the rights and responsibilities of the people and the government in helping to prevent possible problems with the ecology of the earth. This, however, is not an easy thing and students need to see that there are two sides to a problem (sometimes more) and that it may not be as easy as they think to change things. In this area, there are many possible projects but for my third grade class I would like them to research one of a group of air pollution sources and gather information about what it is. After they have done this each student will plan out a poster and title illustrating the problem they have chosen. As an addition to this research they will then compose a letter to a local government official stating the problem they are concerned about and requesting that that person do something about it.

PLANTS AFFECTED BY POLLUTION

While we are still able to breathe, there are many days

when the pollution levels make it difficult. Chemicals and gases in the air that should not be present cause air pollution. Three main environmental problems that affect plants include Acid Rain, Global Warming, and Ozone Depletion.

Acid rain begins as normal rain but when it falls through clouds of pollution the rain changes into a weak acid. Water vapour sometimes mixes with the exhaust from cars, and factory smoke stacks as it moves over the land. This combination of nitric oxide and sulfur dioxide dissolves in the clouds. This acid can sometimes be as strong as vinegar. This acid is strong enough to damage trees, dissolve marble, and kill fish in ponds and lakes. Small lakes and ponds can be helped temporarily by the addition of baking soda or limestone. The best cure, however, is to control the pollution that is the cause. Acid rain can measure 2.0 or lower on the pH scale.

Global Warming or the greenhouse affect is connected to the fact that the earth is a closed ecosystem. The temperatures are kept to a certain level so that plant life can thrive. The earth's temperature is partly the result of the stratosphere that traps gases in the troposphere and stratosphere. Too much heat would make the earth warmer and with hotter temperatures the balance of life would begin to erode. The so-called greenhouse gases ñ carbon dioxide, water vapour, ozone, methane, nitrous oxide, and chlorofluorocarbons - trap heat that is reflected off the earth's surface. Nature tries to balance itself, but the things that people do upset it.

When people burn wood, coal, and oil, more carbon dioxide is released into the air. The carbon dioxide slows down the movement of heat and acts as an insulator. The result is what would happen if the earth were put inside a closed jar. The earth gets increasingly warm from the sun but it cannot cool off. Trees are an important solution to the problem since the trees help to remove carbon out of the air.

Saving the rain forests from destruction would help. Some scientists speculate that if the earth's atmosphere rises a few degrees it could have a devastating affect on the planet. Plant life would be negatively impacted and the polar ice caps would begin to melt. People often confuse the issue of global warming with that of the ozone depletion but they are two separate occurrences though both are major environmental problems.

The ozone is a thin layer of gas that is about 12 miles up in the earth's stratosphere. This gas helps to protect the earth by filtering the sunlight that hits the earth. Without that layer higher than normal levels of the sun's ultraviolet radiation would come down and affect the population. These harmful rays mean more skin cancer, cataracts, and lung problems from increased smog levels. Chemicals like the chlorofluorocarbons in refrigerators and air conditioners harm the ozone layer as they rise into the air.

Again people can help if they change their buying habits. Don't purchase foam products that can damage the ozone when they are burned or put in landfills. Use recyclable paper or plastic cups instead of Styrofoam that is also toxic. Refrigerators and freezers should be kept in good running order and if possible limit use of air conditioning or use a fan instead. Scientists measure Ozone depletion by using satellites. The layer is getting thinner, and over places like Antarctica it is completely gone.

As a culminating activity the class would make a hypermedia presentation about the need to take care of the earth using a student PowerPoint Programme.

Chapter 6

Insecticides

Insecticides are agents of chemical or biological origin that control insects. Control may result from killing the insect or otherwise preventing it from engaging in behaviours deemed destructive. Insecticides may be natural or manmade and are applied to target pests in a myriad of formulations and delivery systems (sprays, baits, slow-release diffusion, etc.). The science of biotechnology has, in recent years, even incorporated bacterial genes coding for insecticidal proteins into various crop plants that deal death to unsuspecting pests that feed on them.

The purpose of this brief chapter is to provide a handshake overview of what insecticides are, and a short background and a review of the major insecticide classes that have been or are used today to cope with insect pests. Though by no means exhaustive, we will touch on major classes and technologies whether decades old or recently revealed.

Some 10,000 species of the more than 1 million species of insects are crop-eating, and of these, approximately 700 species worldwide cause most of the insect damage to man's crops, in the field and in storage.

Humanoids have been on earth for more than 3 million years, while insects have existed for at least 250 million years. We can guess that among the first approaches used by our primitive ancestors to reduce insect annoyance was hugging smoky fires or spreading mud and dust over their skin to repel biting and tickling insects, a practice resembling the habits of elephants, swine, and water buffalo. Today, such approaches would be classed as *repellents,* a category of *insecticides.*

Historians have traced the use of pesticides to the time of Homer around 1000 B.C., but the earliest records of insecticides pertain to the burning of "brimstone" (sulfur) as a fumigant. Pliny the Elder recorded most of the earlier insecticide uses in his *Natural History*. Included among these were the use of gall from a green lizard to protect apples from worms and rot. Later, we find a variety of materials used with questionable results extracts of pepper and tobacco, soapy water, whitewash, vinegar, turpentine, fish oil, brine, lye among many others.

At the beginning of World War II, our insecticide selection was limited to several arsenicals, petroleum oils, nicotine, pyrethrum, rotenone, sulfur, hydrogen cyanide gas, and cryolite. It was World War II that opened the *Modern Era of Chemical* control with the introduction of a new concept of insect control –synthetic organic insecticides, the first of which was DDT.

ORGANOCHLORINES

The organochlorines are insecticides that contain carbon (thus *organo-*), hydrogen, and chlorine. They are also known by other names *chlorinated hydrocarbons, chlorinated organics, chlorinated insecticides,* and *chlorinated synthetics*. The organochlorines are now primarily of historic interest, since few survive in today's arsenal.

DIPHENYL ALIPHATICS

The oldest group of the organochlorines is the *diphenyl aliphatics,* which included DDT, DDD, dicofol, ethylan, chlorobenzilate, and methoxychlor. DDT is probably the best known and most notorious chemical of the 20th century. It is also fascinating, and remains to be acknowledged as the most useful insecticide developed. More than 4 billion pounds of DDT were used throughout the world, beginning in 1940, and in the U.S. ending essentially in 1973, when the U.S. Environmental Protection Agency canceled all uses. The remaining First World countries rapidly followed suit. DDT is still effectively used for malaria control in several third world countries. In 1948, Dr. Paul Muller, a Swiss entomologist, was awarded the Nobel Prize in Medicine for his lifesaving discovery of DDT as an insecticide useful in the control of malaria, yellow fever and many other insect-vectored diseases.

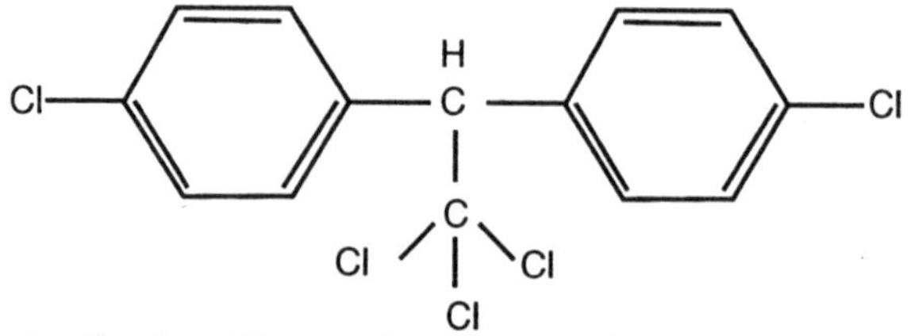

Mode of action–The mode of action for DDT has never been clearly established, but in some complex manner it destroys the delicate balance of sodium and potassium ions within the axons of the neuron in a way that prevents normal transmission of nerve impulses, both in insects and mammals.

It apparently acts on the sodium channel to cause "leakage" of sodium ions. Eventually the affected neurons fire impulses spontaneously, causing the muscles to twitch– "DDT jitters"– followed by convulsions and death. DDT has a negative temperature correlation–the lower the surrounding temperature the more toxic it becomes to insects.

HEXCHLOROCYCLOHEXANE (HCH)

Also known as benzenehexachloride (BHC), the insecticidal properties of HCH were discovered in 1940 by French and British entomologists. In its technical grade, there are five isomers, *alpha, beta, gamma, delta* and *epsilon*. Surprisingly, only the *gamma* isomer has insecticidal properties.

Consequently, the *gamma* isomer was isolated in manufacture and sold as the odorless insecticide *lindane*. In contrast, technical grade HCH has a strong musty odor and flavour, which can be imparted to treated crops and animal products. Because of its very low cost, HCH is still used in many developing countries. In 2002, the U.S. EPA removed all food-related (tolerance-requiring) uses of lindane from the U.S.

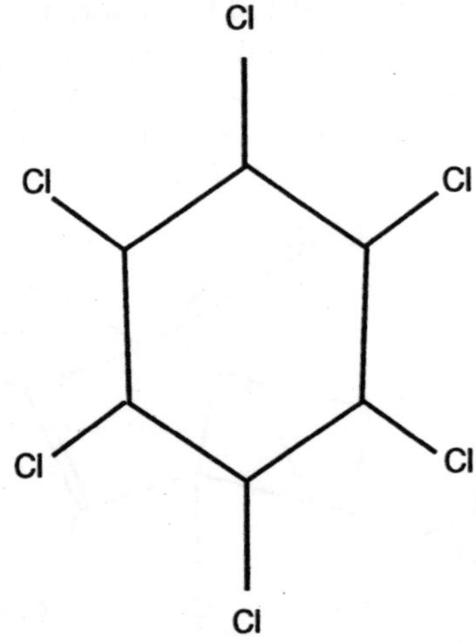

Mode of Action

The effects of HCH superficially resemble those of DDT, but occur much more rapidly, and result in a much higher rate of respiration in insects. The *gamma* isomer is a neurotoxicant whose effects are normally seen within hours as increased activity, tremors, and convulsions leading to prostration. It too, exhibits a negative temperature correlation, but not as pronounced as that of DDT.

CYCLODIENES

The cyclodienes appeared after World War II chlordane, 1945, aldrin and dieldrin, 1948; heptachlor, 1949; endrin, 1951; mirex, 1954; endosulfan, 1956; and chlordecone, 1958. There were other cyclodienes of minor importance developed in the U.S. and Germany. Most of the cyclodienes are persistent insecticides and are stable in soil and relatively stable to the ultraviolet of sunlight. As a result, they were used in greatest quantity as soil insecticides (especially chlordane, heptachlor, aldrin, and dieldrin) for the control of termites and soil-borne insects whose larval stages feed on the roots of plants.

To appreciate the effectiveness of these materials as termiticides, consider that wood and wooden structures treated with chlordane, aldrin, and dieldrin in the year of their development are still protected from damage—after more than 60 years! The cyclodienes were the most effective, long-lasting and economical termiticides ever developed.

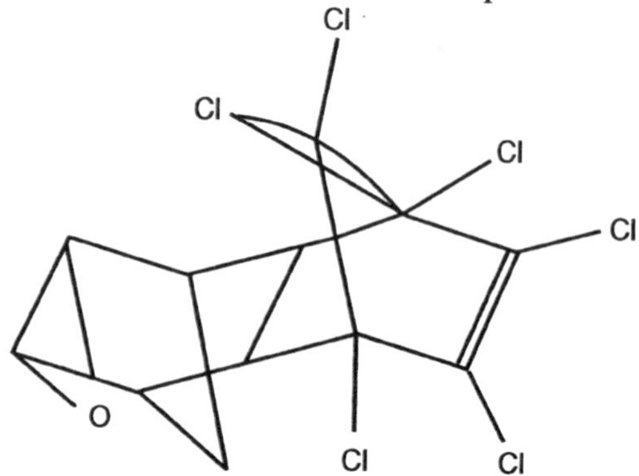

Because of their persistence in the environment, resistance that developed in several soil insect pests, and in some instances *biomagnification* in wildlife food chains, most agricultural uses of cyclodienes were canceled by the EPA between 1975 and 1980, and their use as termiticides canceled in 1984-88.

Mode of Action

Unlike DDT and HCH, the cyclodienes have a positive temperature correlation–their toxicity increases with increasing ambient temperature. Their modes of action are also not clearly understood. However, it is known that this group acts on the inhibitory mechanism called the GABA (g-aminobutyric acid) receptor.

This receptor operates by increasing chloride ion permeability of neurons. Cyclodienes prevent chloride ions from entering the neurons, and thereby antagonize the "calming" effects of GABA. Cyclodienes appear to affect all animals similarly, first with the nervous activity followed by tremors, convulsions and prostration.

POLYCHLOROTERPENES

Only two polychloroterpenes were developed–toxaphene in 1947, and strobane in 1951. Toxaphene had by far the greatest use of any single insecticide in agriculture, while strobane was relatively insignificant.

Toxaphene was used on cotton, first in combination with DDT, for alone it had minimal insecticidal qualities. Then, in 1965, after several major cotton insects became resistant to DDT, toxaphene was formulated with methyl parathion, an organophosphate insecticide mentioned later.

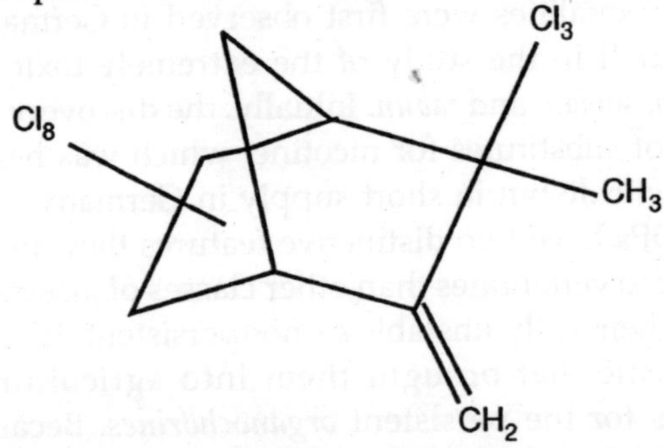

Toxaphene is a mixture of more than 177 10-carbon polychlorinated derivatives. These materials persist in the soil, though not as long as the cyclodienes, and disappear from the surfaces of plants in 3-4 weeks. This disappearance is attributed more to volatility than to photolysis or plant metabolism.

Toxaphene is rather easily metabolized by mammals and birds, and is not stored in body fat nearly to the extent of DDT, HCH and the cyclodienes. Despite its low toxicity to insects, mammals and birds, fish are highly susceptible to toxaphene poisoning, in the same order of magnitude as to the cyclodienes. Toxaphene's registrations were canceled by EPA in 1983.

Mode of Action

Toxaphene and strobane act on the neurons, causing an imbalance in sodium and potassium ions, similar to that of the cyclodiene insecticides.

ORGANOPHOSPHATES

Organophosphates (OPs) is the term that includes all insecticides containing phosphorus. Other names used, but no longer in vogue, are *organic phosphates, phosphorus insecticides, nerve gas relatives,* and *phosphoric acid esters*. All organophosphates are derived from one of the phosphorus acids, and as a class are generally the most toxic of all pesticides to vertebrates.

Because of the similarity of OP chemical structures to the "nerve gases," their modes of action are also similar. Their insecticidal qualities were first observed in Germany during World War II in the study of the extremely toxic OP nerve gases *sarin, soman,* and *tabun*. Initially, the discovery was made in search of substitutes for nicotine, which was heavily used as an insecticide but in short supply in Germany.

The OPs have two distinctive features they are generally more toxic to vertebrates than other classes of insecticides, and most are chemically unstable or nonpersistent. It is this latter characteristic that brought them into agricultural use as substitutes for the persistent *organochorines*. Because of the relatively high toxicity of the OP's, EPA, under provisions of the Food Quality Protection Act (1996), undertook an extensive reappraisal of the entire class beginning in the late 1990's. Many OP's were voluntarily canceled and others lost uses.

Mode of action–The OPs work by inhibiting certain important enzymes of the nervous system, namely *cholinesterase* (ChE). The enzyme is said to be *phosphorylated* when it becomes attached to the phosphorous moiety of the insecticide, a binding that is irreversible. This inhibition results in the accumulation of acetylcholine (ACh) at the neuron/neuron and neuron/muscle (neuromuscular) junctions or synapses, causing rapid twitching of voluntary muscles and finally paralysis.

CLASSIFICATION

All OPs are esters of phosphorus having varying combinations of oxygen, carbon, sulfur and nitrogen attached, resulting in six different subclasses phosphates, phospho-nates, phosphorothioates, phosphorodithioates, phosphorothiolates and phosphoramidates. These subclasses are easily identified by their chemical names.

The OPs are generally divided into three groups–*aliphatic, phenyl,* and *heterocyclic* derivatives.

ALIPHATICS

The aliphatic OPs are carbon chain-like in structure. The first OP brought to agriculture, TEPP belonged to this group. Other examples are malathion, trichlorfon, monocrotophos, dimethoate, oxydemetonmethyl, dicrotophos, disulfoton, dichlorvos, mevinphos, methamidophos, and acephate.

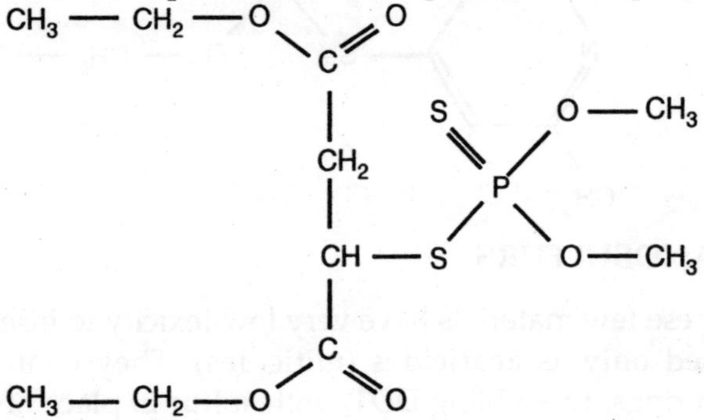

CH_3 — CH_2 — O O

C

S O — CH_3

CH_2

P

CH — S O — CH_3

C

CH_3 — CH_2 — O O

PHENYL DERIVATIVES

The phenyl OPs contain a phenyl ring with one of the ring hydrogens displaced by attachment to the phosphorus moiety and other hydrogens frequently displaced by Cl, NO_2, CH_3, CN, or S. The phenyl OPs are generally more stable than the aliphatics, thus their residues are longer lasting. The first phenyl OP brought into agriculture was parathion in 1947. Examples of other phenyl OPs are methyl parathion, profenofos, sulprofos, isofenphos, fenitrothion, fenthion, and famphur.

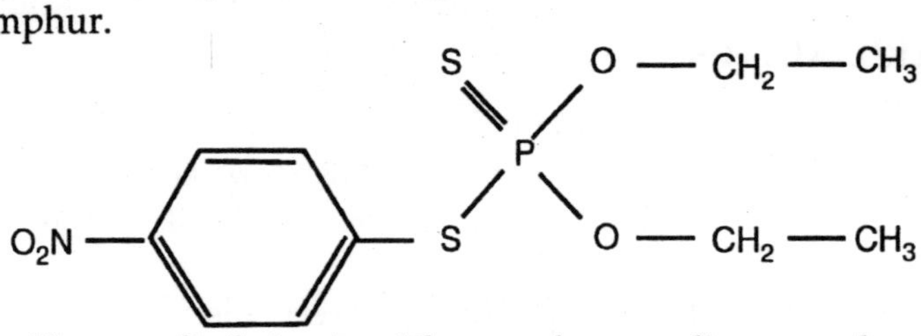

S O — CH_2 — CH_3

P

O_2N — S O — CH_2 — CH_3

Heterocyclic derivatives–The term *heterocyclic* means that the ring structures are composed of different or unlike atoms, e.g.,oxygen, nitrogen or sulfur. The first of this group was diazinon introduced in 1952. Other examples in this group are azinphos-methyl, azinphos-ethyl, chlorpyrifos, methidathion, phosmet, isazophos, and chlorpyrifos-methyl.

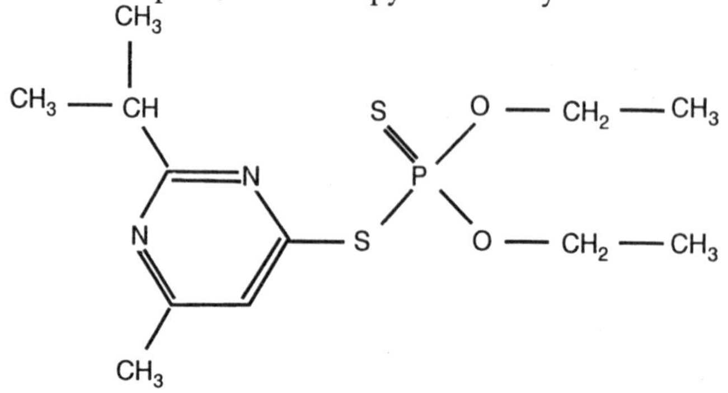

CH_3, CH_3—CH, S, P, O—CH_2—CH_3, N, N, S, O—CH_2—CH_3, CH_3

ORGANOSULFURS

These few materials have very low toxicity to insects and are used only as acaricides (miticides). They contain two phenyl rings, resembling DDT, with sulfur in place of carbon as the central atom. These include tetradifon, propargite, and ovex.

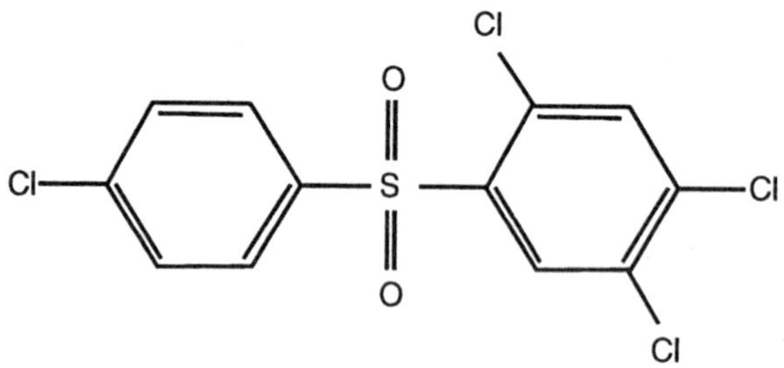

Cl, O, Cl—, S, Cl, O, Cl

CARBAMATES

The carbamate insecticides are derivatives of carbamic acid (as the OPs are derivatives of phosphoric acid). And like the OPs, their mode of action is that of inhibiting the vital enzyme *cholinesterase* (ChE).

The first successful carbamate insecticide, carbaryl, was introduced in 1956. More of it has been used worldwide than all the remaining carbamates combined. Two distinct qualities have made it the most popular carbamate its very low mammalian oral and dermal toxicity and an exceptionally broad spectrum of insect control. Other long-standing carbamate insecticides are methomyl, carbofuran, aldicarb, oxamyl, thiodicarb, methiocarb, propoxur, bendiocarb, carbosulfan, aldoxycarb, promecarb, and fenoxycarb. Carbamates more recently introduced include primicarb, indoxacarb, alanycarb and furathiocarb.

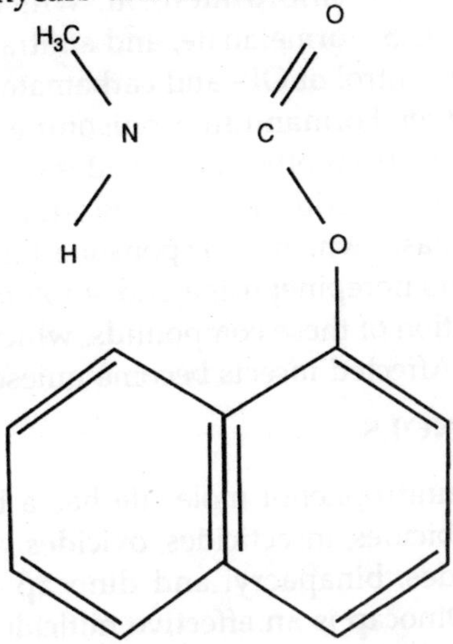

Mode of Action

Carbamates inhibit cholinesterase (ChE) as OPs do, and they behave in almost identical manner in biological systems, but with two main differences. Some carbamates are potent inhibitors of aliesterase (miscellaneous aliphatic esterases whose exact functions are not known), and their selectivity is sometimes more pronounced against the ChE of different species. Second, ChE inhibition by carbamates is reversible. When ChE is inhibited by a carbamate, it is said to be *carbamylated,* as when an OP results in the enzyme being *phosphorylated.*

In insects, the effects of OPs and carbamates are primarily those of poisoning of the central nervous system, since the insect neuromuscular junction is not cholinergic, as in mammals. The only cholinergic synapses known in insects are in the central nervous system. (The chemical neuromuscular junction transmitter in insects is thought to be glutamic acid.)

FORMAMIDINES

The formamidines comprise a small group of insecticides. Three examples are chlordimeform, which is no longer registered in the U.S., formetanate, and amitraz. Their current value lies in the control of OP- and carbamate-resistant pests.

Mode of action–Formamidine poisoning symptoms are distinctly different from other insecticides.

Their proposed action is the inhibition of the enzyme monoamine oxidase, which is responsible for degrading the neurotransmitters norepinephrine and serotonin. This results in the accumulation of these compounds, which are known as *biogenic amines*. Affected insects become quiescent and die.

DINITROPHENOLS

The basic dinitrophenol molecule has a broad range of toxicities–as herbicides, insecticides, ovicides, and fungicides. Of the insecticides, binapacryl and dinocap were the most recently used. Dinocap is an effective miticide and was very heavily used as a fungicide for the control of powdery mildew fungi. Because of the inherent toxicity of the dinitrophenols, they have all been withdrawn.

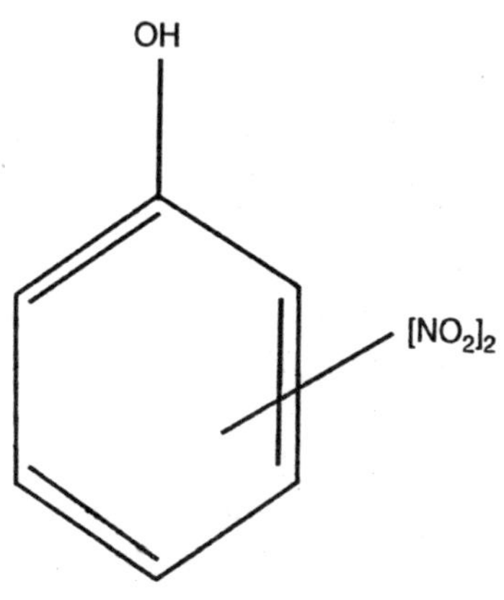

Mode of action–Dinitrophenols act by uncoupling or inhibiting oxidative phosphorylation, which basically prevents the formation of the high-energy phosphate molecule, adenosine triphosphate (ATP).

ORGANOTINS

The organotins are a group of acaricides that double as fungicides. Of particular interest is cyhexatin, one of the most selective acaricides known, introduced in 1967. Fenbutatin-oxide has been used extensively against mites on deciduous fruits, citrus, greenhouse crops, and ornamentals.

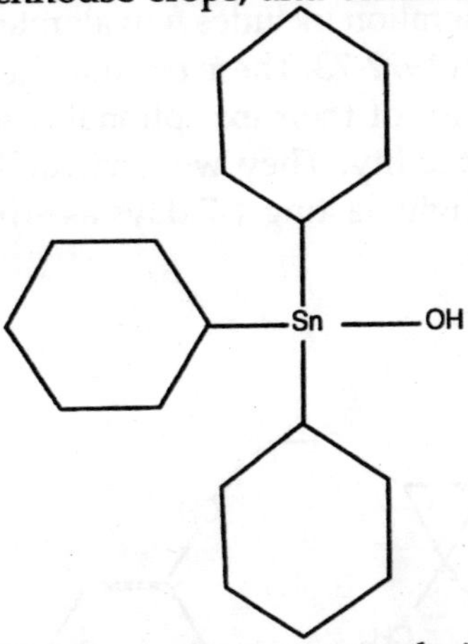

Sn —— OH

Mode of action–These tin compounds inhibit oxidative phosphorylation at the site of dinitrophenol uncoupling, preventing the formation of the high-energy phosphate molecule adenosine triphosphate (ATP). These trialkyl tins also inhibit photophosphorylation in chloroplasts, the chlorophyll-bearing subcellular units) and could therefore serve as algicides.

PYRETHROIDS

Natural pyrethrum has seldom been used for agricultural purposes because of its cost and instability in sunlight. In recent decades, many synthetic pyrethrin-like materials have become available. They were originally referred to as *synthetic pyrethroids*. Currently, the better nomenclature is simply *pyrethroids*. These are stable in sunlight and are generally effective against most agricultural insect pests when used at the very low rates of 0.01 to 0.1 pound per acre.

The pyrethroids have an interesting evolution, which is conveniently divided into four generations. The first generation contains only one pyrethroid, allethrin, which appeared in 1949. Its synthesis was very complex, involving 22 chemical reactions to reach the final product. The second generation includes tetramethrin, followed by resmethrin in 1967 (20X as effective as pyrethrum), then bioresmethrin, then Bioallethrin, and finally phonothrin.

The third generation includes fenvalerate, and permethrin which appeared in 1972-73. These became the first agricultural pyrethroids because of their exceptional insecticidal activity and their photostability. They were virtually unaffected by ultraviolet in sunlight, lasting 4-7 days as efficacious residues on crop foliage.

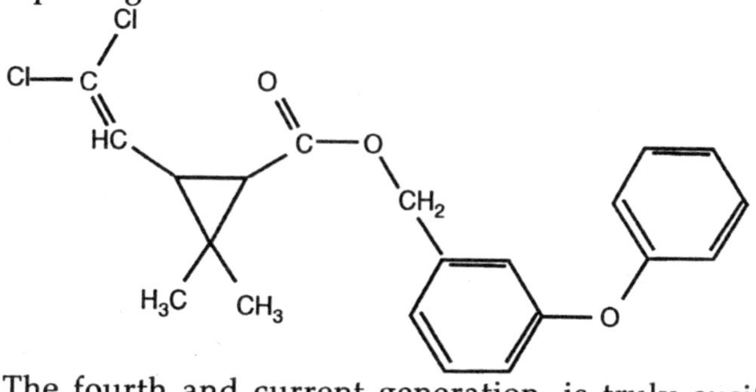

The fourth and current generation, is truly exciting because of their effectiveness in the range of 0.01 to 0.05 lb ai/ A. These include bifenthrin, *lambda*-cyhalothrin, cypermethrin, cyfluthrin, deltamethrin, fenpropathrin, flucythrinate, fluvalinate, prallethrin, *tau*-fluvalinate tefluthrin, tralomethrin, and *zeta*-cypermethrin.

All of these are photostable, that is, they do not undergo photolysis (splitting) in sunlight. And because they have minimal volatility they provide extended residual effectiveness, up to 10 days under optimum conditions.

Recent additions to the fourth generation pyrethroids are acrinathrin, imiprothrin, registered in 1998, and *gamma*-

cyhalothrim, which is in development.

Mode of Action

The pyrethroids share similar modes of action, resembling that of DDT, and are considered axonic poisons. They apparently work by keeping open the sodium channels in neuronal membranes. There are two types of pyrethroids. Type I, among other physiological responses, have a negative temperature coefficient, resembling that of DDT. Type II, in contrast have a positive temperature coefficient, showing increased kill with increase in ambient temperature.

Pyrethroids affect both the peripheral and central nervous system of the insect. They initially stimulate nerve cells to produce repetitive discharges and eventually cause paralysis. Such effects are caused by their action on the sodium channel, a tiny hole through which sodium ions are permitted to enter the axon to cause excitation. The stimulating effect of pyrethroids is much more pronounced than that of DDT.

NICOTINOIDS

The nicotinoids are a newer class of insecticides with a new mode of action. They have been previously referred to as *nitro-quanidines, neonicotinyls, neonicotinoids, chloronicotines,* and more recently as the *chloronicotinyls.* Just as the synthetic pyrethroids are similar to and modeled after the natural pyrethrins, so too, are the nicotinoids similar to and modeled after the natural nicotine. Imidacloprid was introduced in Europe and Japan in 1990 and first registered in the U.S. in 1992. It is currently marketed as several proprietary products worldwide, e.g., Admire, Confidor, Gaucho, Merit, Premier, Premise and Provado. Very possibly it is used in the greatest volume globally of all insecticides.

Imidacloprid is a systemic insecticide, having good root-systemic characteristics and notable contact and stomach action. It is used as a soil, seed or foliar treatment in cotton, rice cereals, peanuts, potatoes, vegetables, pome fruits, pecans and turf, for the control of sucking insects, soil insects, whiteflies, termites, turf insects and the Colourado potato beetle, with long residual control. Imidacloprid has no effect on mites or nematodes.

Cl N NO$_2$ N N_2C N N—H

Other nicotinoids include acetamiprid, thiamethoxam, nitenpyram, clothianidin, dinotefuran and thiacloprid. U.S. registrations for acetamiprid, thiamethoxam and thiacloprid were granted in 2002 and clothianidin in 2003.

Mode of Action

The nicotinoids act on the central nervous system of insects, causing irreversible blockage of postsynaptic nicotinergic acetylcholine receptors.

SPINOSYNS

Spinosyns are among the newest classes of insecticides, represented by spinosad. Spinosad is a fermentation metabolite of the actinomycete *Saccharopolyspora spinosa*, a soil-inhabiting microorganism. It has a novel molecular structure and mode of action that provide excellent crop protection typically associated with synthetic insecticides, first registered for use on cotton in 1997.

Spinosad is a mixture of spinosyns A and D. It is particularly effective as a broad-spectrum material for most caterpillar pests at the astonishing rates of 0.04 to 0.09 pound of active ingredient (18 to 40 grams) per acre. It has both contact and stomach activity against lepidopteran larvae, leaf miners, thrips, and termites, with long residual activity. Crops registered include cotton, vegetables, tree fruits, ornamentals and others.

Mode of action–Spinosad acts by disrupting binding of

acetylcholine in nicotinic acetylcholine receptors at the postsynaptic cell.

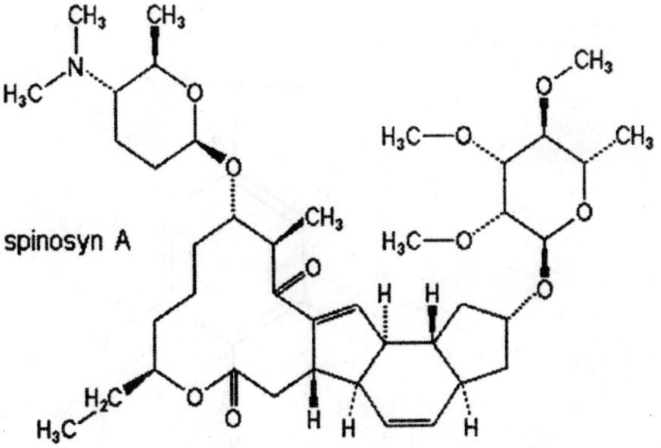

spinosyn A

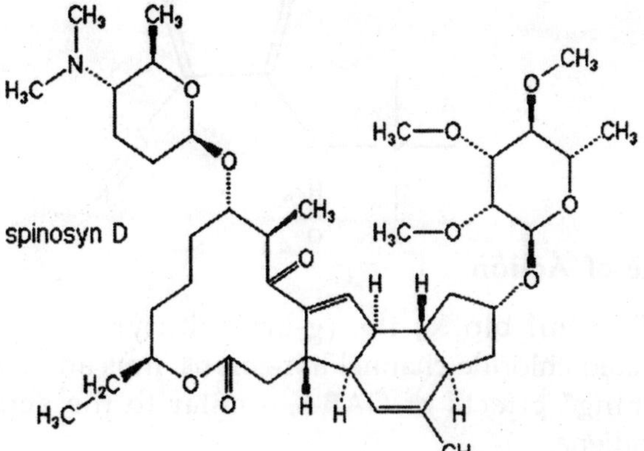

spinosyn D

FIPROLES

Fipronil is the only insecticide in this new class, introduced in 1990 and registered in the U.S. in 1996. It is a systemic material with contact and stomach activity. Fipronil is used for the control of many soil and foliar insects, (e.g., corn rootworm, Colourado potato beetle, and rice water weevil) on a variety of crops, primarily corn, turf, and for public health insect control.

It is also used for seed treatment and formulated as baits for cockroaches, ants and termites. Fipronil is effective against insects resistant or tolerant to pyrethroid, organophosphate and carbamate insecticides.

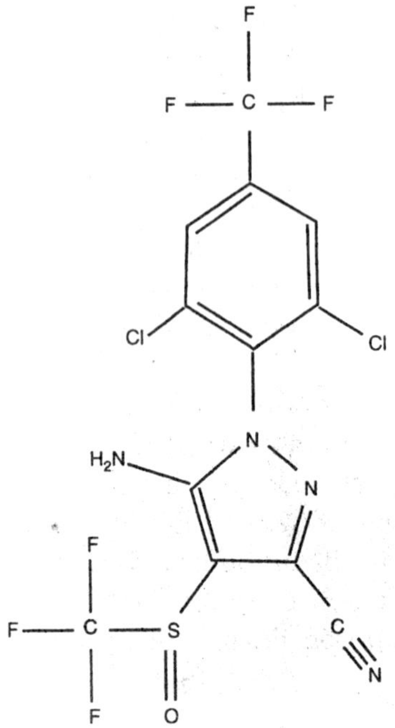

Mode of Action

Fipronil blocks the (g-aminobutyric acid- (GABA) regulated chloride channel in neurons, thus antagonizing the "calming" effects of GABA, similar to the action of the Cyclodienes.

PYRROLES

Chlorfenapyr is the first and only member of this unique chemical group, as both a contact and stomach insecticide-miticide. It is used on cotton and experimentally on corn, soybeans, vegetables, tree and vine crops, and ornamentals to control whitefly, thrips, caterpillars, mites, leafminers, aphids, and Colourado potato beetle.

It has ovicidal activity on some species. EPA took the unusual step of refusing to register chlorfenapyr in 2000 for cotton insect control because of potential hazards to birds. However, labels for greenhouse ornamentals were granted in 2001.

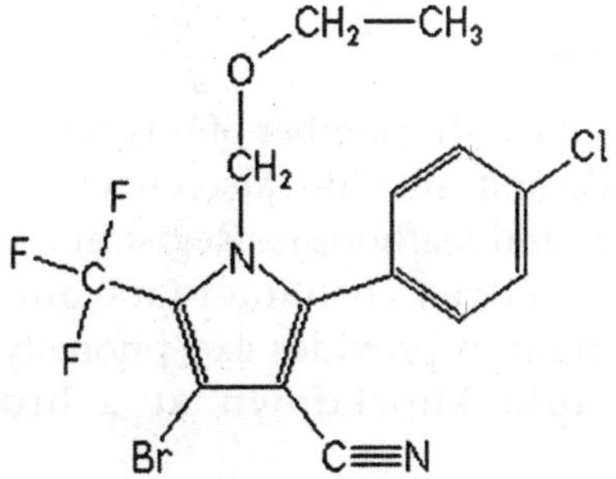

Mode of Action

Chlorfenapyr is an "uncoupler" or inhibitor of oxidative phosphorylation, preventing the formation of the crucial energy molecule adenosine triphosphate (ATP).

PYRAZOLES

The original pyrazoles were tebufenpyrad and fenpyroximate (not illustrated). These were designed primarily as non-systemic contact and stomach miticides, but do have limited effectiveness on psylla, aphids, whitefly, and thrips. Tebufenpyrad, registered by EPA in 2002, is used on cotton, soybeans, vegetables, pome fruits, grapes and citrus. Fenpyroximate controls all stages of mites, gives fast knockdown, inhibits molting of immature stages of mites, and has long residual activity. Newer members of this class include ethiprole which is active on a broad sprectum of chewing and sucking insects, and tolfenpyrad which is reputed to active on pests infesting cole and cucurbit crops.

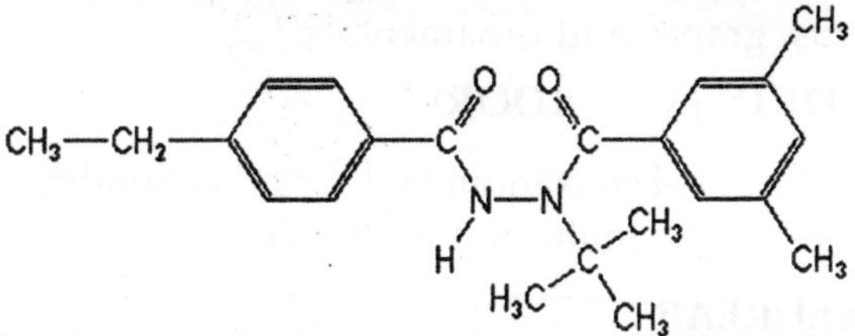

Mode of action–Their mode of action is that of inhibiting mitochondrial electron transport at the NADH-CoQ reductase site, leading to the disruption of adenosine triphosphate (ATP) formation, the crucial energy molecule.

PYRIDAZINONES

Pyridaben is the only member of this class. It is a selective contact insecticide and miticide, also effective against thrips, aphids, whiteflies and leafhopprs. Registrations are for pome fruits, almonds, citrus, ornamentals and greenhouse ornamentals. Pyridaben provides exceptionally long residual control, and rapid knockdown at a broad range of temperatures.

Mode of Action

Pyridaben is a metabolic inhibitor that interrupts mitochondrial electron transport at Site 1, similar to the Quinazolines, below.

H_3C CH_3 N N C CH_3 CH_3 CH_3—C—CH_2—S O CH_3 Cl

QUINAZOLINES

The quinazolines offer a unique chemical configuration, consisting only of one insecticide, fenazaquin. Fenazaquin is a contact and stomach miticide. It has ovicidal activity, gives rapid knockdown, and controls all stages of mites. Not yet registered in the U.S., it is used on cotton, stone and pome fruits, citrus, grapes and ornamentals.

FENAZAQUIN (MATADOR)

Mode of action–Fenazaquin inhibits mitochondrial electron transport at Site 1, similar to the Pyridazinones, above.

BENZOYLUREAS

Benzoylureas are an entirely different class of insecticides that act as insect growth regulators (IGRs). Rather than being the typical poisons that attack the insect nervous system, they interfere with chitin synthesis and are taken up more by ingestion than by contact. Their greatest value is in the control of caterpillars and beetle larvae.

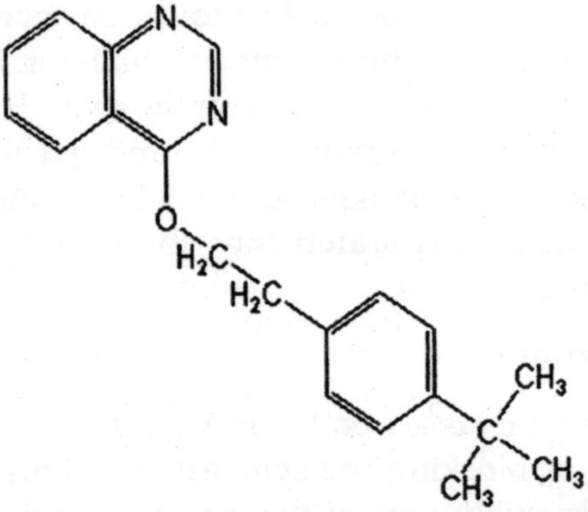

Benzoylureas were first used in Central America in 1985, to control a severe, resistant leafworm complex (*Spodoptera, Trichoplusia*) outbreak in cotton. The withdrawal of the ovicide chlordimeform made their control quite difficult due to their high resistance to almost all insecticide classes, including the pyrethroids. The benzoylureas were introduced in 1978 by Bayer of Germany, triflumuron being the first. Others appearing since then are chlorfluazuron, followed by teflubenzuron, hexaflumuron, flufenoxuron, and flucycloxuron. Others are flurazuron, novaluron, and diafenthiuron, bistrifluron and noviflumuron. Until recently lufenuron was the newest addition to this group, appearing in 1990. Among the newer benzoylureas only hexaflumuron and novaluron have been registered by EPA.

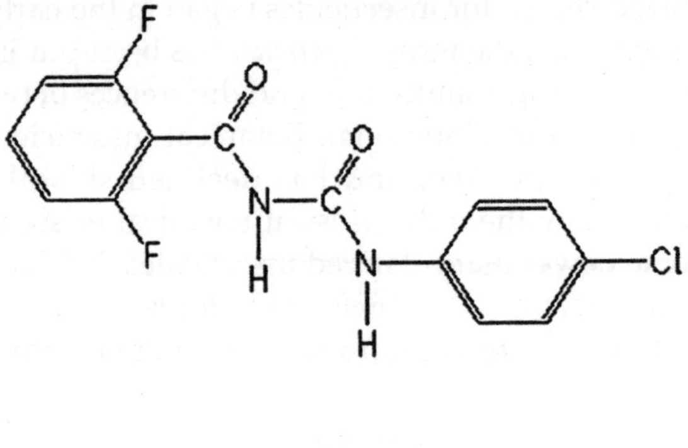

The only other benzoylurea registered in the U.S. is diflubenzuron. It was first registered in 1982 for gypsy moth, cotton boll weevil, most forest caterpillars, soybean caterpillars, and mushroom flies, but now with a much broader range of registrations. Though not a benzoylurea, cyromazine, a triazine, is also a potent chitin synthesis inhibitor. It is selective toward Dipterous species and used for the control of leafminers in vegetable crops and ornamentals, and fed to poultry or sprayed to control flies in manure of broiler and egg producing operations, and incorporated into compost of mushroom houses for fungus gnats.

Mode of Action

The benzoylureas act on the larval stages of most insects by inhibiting or blocking the synthesis of chitin, a vital and almost indestructible part of the insect exoskeleton. Typical effects on developing larvae are the rupture of malformed cuticle or death by starvation. Adult female boll weevils exposed to diflubenzuron lay eggs that do not hatch. And, mosquito larvae control can be achieved with as little as 1.0 gram of diflubenzuron per acre of surface water.

BOTANICALS

Botanical insecticides are of great interest to many, for they are *natural* insecticides, toxicants derived from plants. Historically, the plant materials have been in use longer than any other group, with the possible exception of sulfur. Tobacco, pyrethrum, derris, hellebore, quassia, camphor, and turpentine were some of the more important plant products in use before the organized search for insecticides began in the early 1940s.

In recent years the term *biorational* has been put into play by the EPA. There are similarities and differences between the terms botanical and biorational. Botanical insecticide use in the U.S. peaked in 1966, and has declined steadily since. Pyrethrum is now the only classical botanical of significance in use. Some newer plant-derived insecticides that have come into use are referred to as *florals or scented plant chemicals* and include, among others, limonene, cinnamaldehyde and eugenol. In addition, there is azadirachtin from the neem tree which is used in greenhouse and on ornamentals.

PYRETHRUM

Pyrethrum is extracted from the flowers of a chrysanthemum grown in Kenya and Ecuador. It is one of the oldest and safest insecticides available. The ground, dried flowers were used in the early 19th century as the original louse powder to control body lice in the Napoleonic Wars. Pyrethrum acts on insects with phenomenal speed causing immediate paralysis, thus its popularity in fast knockdown household aerosols. However, unless it is formulated with one of the *synergists,* most of the paralyzed insects recover to once again become pests. Pyrethrum is a mixture of four compounds pyrethrins I and II and cinerins I and II.

Mode of action–Pyrethrum is an axonic poison, as are the synthetic pyrethroids and DDT. Axonic poisons are those that in some way affect the electrical impulse transmission along the axons, the elongated extensions of the neuron cell body. Pyrethrum and some pyrethroids have a greater insecticidal effect when the temperature is lowered, a negative temperature coefficient, as does DDT.

They affect both the peripheral and central nervous system of the insect. Pyrethrum initially stimulates nerve cells to produce repetitive discharges, leading eventually to paralysis. Such effects are caused by their action on the sodium channel, a tiny hole through which sodium ions are permitted to enter the axon to cause excitation. These effects are produced in insect nerve cord, which contains ganglia and synapses, as well as in giant nerve fibre axons.

NICOTINE

Nicotine is extracted by several methods from tobacco, and is effective against most all types of insect pests, but is used particularly for aphids and caterpillars–soft bodied insects. Nicotine is an alkaloid, a chemical class of heterocyclic compounds containing nitrogen and having prominent physiological properties. Other well-known alkaloids that are not insecticides are caffeine (coffee, tea), quinine (cinchona bark), morphine (opium poppy), cocaine (coca leaves), ricinine (a poison in castor oil beans), strychnine, coniine (spotted hemlock, the poison used by Socrates), and, finally LSD (a hallucinogen from the ergot fungus attacking grain).

Mode of Action

Nicotine action is one of the first, classic modes of action identified by pharmacologists. Drugs that act similarly to nicotine are said to have a nicotinic response. Nicotine mimics acetylcholine (ACh) at the neuromuscular (nerve/muscle) junction in mammals, and results in twitching, convulsions, and death, all in rapid order. In insects the same action is observed, but only in the central nervous system ganglia.

ROTENONE

Rotenone or rotenoids are produced in the roots of two genera of the legume family *Derris* and *Lonchocarpus* (also called cubé) grown in South America. It is both a stomach and contact insecticide and used for the last century and a half to control leaf-eating caterpillars, and three centuries prior to that in South America to paralyze fish, causing them to surface and be easily captured. Today, rotenone is used in the same way to reclaim lakes for game fishing. Used on a prescribed basis, it eliminates all fish, closing the lake to reintroduction of rough species. It is a selective piscicide in that it kills all fish at dosages that are relatively nontoxic to fish food organisms, and is degraded rapidly.

Mode of Action

Rotenone is a respiratory enzyme inhibitor, acting between NAD+ (a coenzyme involved in oxidation and reduction in metabolic pathways) and coenzyme Q (a respiratory enzyme responsible for carrying electrons in some electron transport chains), resulting in failure of the respiratory functions.

LIMONENE OR *D*-LIMONENE

Limonene or *d*-Limonene is the latest addition to the

botanicals. Limonene belongs to a group often called *florals* or *scented plant chemicals*. Extracted from citrus peel, it is effective against all external pests of pets, including fleas, lice, mites, and ticks, and is virtually nontoxic to warm-blooded animals. Several insecticidal substances occur in citrus oil, but the most important is limonene, which constitutes about 98% of the orange peel oil by weight.

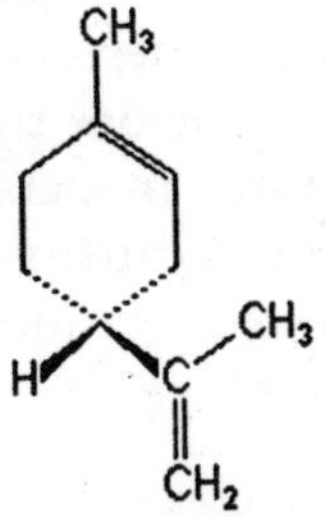

Two other recently introduced floral products are eugenol (Oil of Cloves) and cinnamaldehyde (derived from Ceylon and Chinese cinnamon oils). They are used on ornamentals and many crops to control various insects.

Mode of Action

Its mode of action is similar to that of pyrethrum. It affects the sensory nerves of the peripheral nervous system, but it is not a ChE inhibitor.

NEEM

Neem oil extracts are squeezed from the seeds of the neem tree and contain the active ingredient *azadirachtin*, a nortriterpenoid belonging to the lemonoids. Azadirachtin has shown some rather sensational insecticidal, fungicidal and bactericidal properties, including insect growth regulating qualities. Azatin is marketed as an insect growth regulator, and Align-and Nemix as a stomach/contact insecticide for greenhouse and ornamentals.

Mode of Action

Azadirachtin disrupts molting by inhibiting biosynthesis or metabolism of ecdysone, the juvenile molting hormone.

Synergists or Activators

Synergists are not in themselves considered toxic or insecticidal, but are materials used with insecticides to synergize or enhance the activity of the insecticides. The first was introduced in 1940 to increase the effectiveness of pyrethrum. Since then many materials have appeared, but only a few are still marketed. Synergists are found in most all household, livestock and pet aerosols to enhance the action of the fast knockdown insecticides pyrethrum, allethrin, and resmethrin, against flying insects. Current synergists, such as piperonyl butoxide, contain the methylenedioxyphenyl moiety, a molecule found in sesame oil and later named *sesamin*.

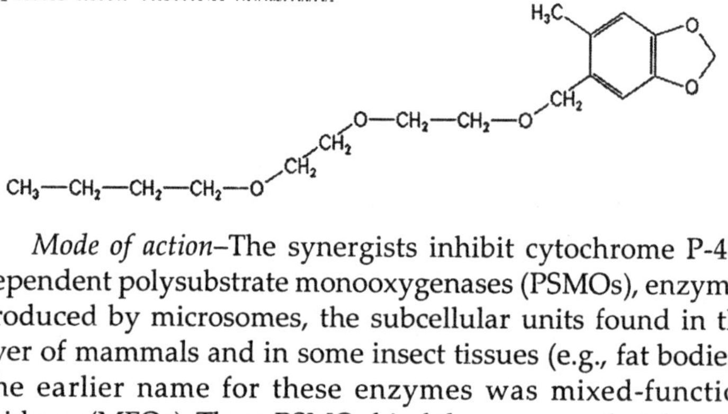

Mode of action–The synergists inhibit cytochrome P-450 dependent polysubstrate monooxygenases (PSMOs), enzymes produced by microsomes, the subcellular units found in the liver of mammals and in some insect tissues (e.g., fat bodies). The earlier name for these enzymes was mixed-function oxidases (MFOs). These PSMOs bind the enzymes that degrade selected foreign substances, such as pyrethrum, allethrin, resmethrin or any other synergized compound. Synergists simply bind the oxidative enzymes and prevent them from degrading the toxicant.

ANTIBIOTICS

In this category belong the *avermectins*, which are insecticidal, acaricidal, and antihelminthic agents that have been isolated from the fermentation products of *Streptomyces avermitilis*, a member of the actinomycete family.

Abamectin is the common name assigned to the avermectins, a mixture of containing 80% avermectin B1*a* and

20% B1*b*, homologs that have about equal biological activity. Clinch is a fire ant bait, and Avid is applied as a miticide/ insecticide. Abamectin has certain local systemic qualities, permitting it to kill mites on a leaf's underside when only the upper surface is treated. The most promising uses for these materials are the control of spider mites, leafminers and other difficult-to-control greenhouse pests, and internal parasites of domestic animals.

Emamectin benzoate is an analog of abamectin, produced by the same fermentation system as abamectin. It was first registered in 1999. It is both a stomach and contact insecticide used primarily for control of caterpillars at the rate of 0.0075 to 0.015 lb (3.5 to 7.0 grams) a.i. per acre. Shortly after exposure, larvae stop feeding and become irreversibly paralyzed, dying in 3-4 days. Rapid photodegradation of both abamectin and emamectin occurs on the leaf surface. More recently, milabectin has been introduced. It is a miticide with activity on piercing/sucking insects and is pending for registration.

Mode of action–Avermectins block the neurotransmitter (g-aminobutyric acid (GABA) at the neuromuscular junction in insects and mites. Visible activity, such as feeding and egg laying, stops shortly after exposure, though death may not occur for several days.

FUMIGANTS

The fumigants are small, volatile, organic molecules that become gases at temperatures above 40°F. They are usually heavier than air and commonly contain one or more of the halogens (Cl, Br, or F). Most are highly penetrating, reaching through large masses of material. They are used to kill insects, insect eggs, nematodes, and certain microorganisms in buildings, warehouses, grain elevators, soils, and greenhouses and in packaged products such as dried fruits, beans, grain, and breakfast cereals.

Although its use is now in decline because of environmental concerns, methyl bromide is the most heavily used of the fumigants, 68,424 metric tons worldwide in 1996, almost half of which is used in the U.S. The dominant use is for preplanting soil treatments, which accounted for 70% of that global total. Quarantine uses account for 5-8%, while 8% is used to treat perishable products, such as flowers and fruits, and 12% for nonperishable products, like nuts and timber. Approximately 6% is used for structural applications, as in drywood termite fumigation of infested buildings.

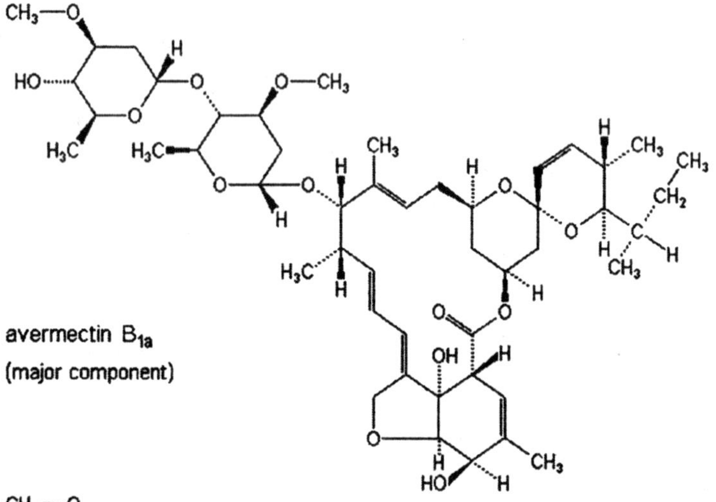

avermectin B_{1a}
(major component)

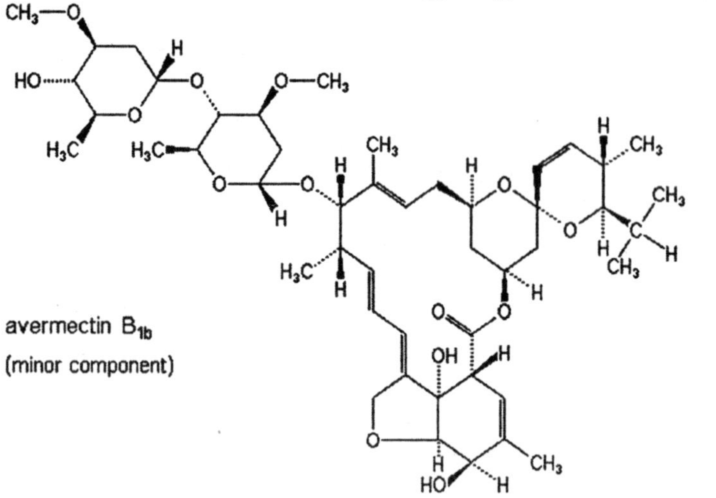

avermectin B_{1b}
(minor component)

With the recently passed change to the Clean Air Act amendments of 1990, U.S. production and importation must be reduced 25% from 1991 levels by 1999. A 50% reduction must be achieved by 2001, followed by a 70% reduction in 2003, and a full ban of the product in 2005. Under the Montreal Protocol, developing countries have until 2015 to phase out methyl bromide production.

Some of the other common fumigants are ethylene dichloride, hydrogen cyanide, sulfuryl fluoride, Vapam, TeloneII, D-D, chlorothene, ethylene oxide, and the familiar home-use moth repellents napthalene crystals and paradichlorobenzene crystals.

Phosphine gas (PH_3) has also replaced methyl bromide in a few applications, primarily for insect pests of grain and food commodities. Treatment requires the use of aluminum or magnesium phosphide pellets, which react with atmospheric moisture to produce the gas. Phosphine, however, is very damaging to fresh commodities and is highly adsorbed onto oil, thus does not perform as a soil fumigant.

Alternatives that can fully replace methyl bromide are unlikely to be available by the deadlines set for replacement. Its low cost and utility on a wide variety of pests are hard to match. Because the loss of MeBr has considerable economic consequences, EPA has made it a priority to find and register replacements. To this end some progress has been made. The chemical 1,3-dichloropropane was registered in 2001 for preplant soil fumigation in strawberries and tomatoes. Moreover, iodomethane and metam-potassium are both being evaluated as soil fumigants.

Mode of action–Fumigants, as a group, are narcotics. That is, they act through means more physical than physiological. The fumigants are liposoluble (fat soluble); they have common symptomology; their effects are reversible; and their activity is altered very little by structural changes in their molecules. As narcotics, they induce narcosis, sleep, or unconsciousness, which in effect is their action on insects.

Liposolubility appears to be an important factor in their action, since these narcotics lodge in lipid-containing tissues found throughout the insect body, including their nervous system.

INSECT REPELLENTS

Historically, repellents have included smoke, plants hung in dwellings or rubbed on the skin as the fresh plant or its brews, oils, pitches, tars, and various earths applied to the body. Before a more edified approach to insect olfaction and behaviour was developed, it was wrongly assumed that if a substance was repugnant to humans it would likewise be repellent to annoying insects.

In recent history, the repellents have been dimethyl phthalate, Indalone, Rutgers 612, dibutyl phthalate, various MGK repellents, benzyl benzoate, the military clothing repellent (N-butyl acetanilide), dimethyl carbate and diethyl toluamide. Of these, only DEET has survived, and is used worldwide for biting flies and mosquitoes. Most of the others have lost their registrations and are no longer available. In 1999, EPA has registered a new insect repellent, N-methylneodecanamide. Rather than being used on humans to repel insects, it is applied to household floors and other surfaces to repel cockroaches and ants.

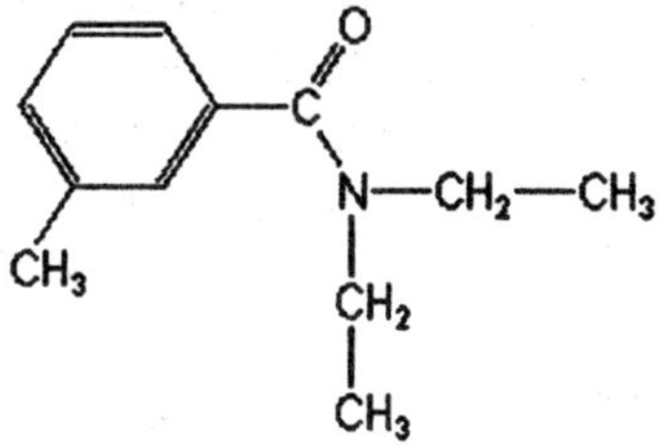

INORGANICS

Inorganic insecticides are those that do not contain carbon. Usually they are white crystals in their natural state, resembling the salts. They are stable chemicals, do not evapourate, and are usually water soluble. Sulfur, mentioned in the introduction, is very likely the oldest known, effective insecticide. Sulfur and sulfur candles were burned by our great-grandparents for every conceivable purpose, from bedbug fumigation to the cleansing of a house just removed from quarantine of smallpox. Today, sulfur is a highly useful material in integrated pest management programs where target pests specificity is important. Sulfur dusts are especially toxic to mites of every variety, such as chiggers and spider mites, and to thrips and newly-hatched scale insects. Sulfur dusts and sprays are also fungicidal, particularly against powdery mildews. Several other inorganic compounds have been used as insecticides mercury, boron, thallium, arsenic, antimony, selenium, and fluoride. Arsenicals have included the copper arsenate, Paris green, lead arsenate, and calcium arsenate. The arsenicals uncouple oxidative phosphorylation, inhibit certain enzymes that contain sulfhydryl (-SH) groups, and coagulate protein by causing the shape or configuration of proteins to change. The inorganic fluorides were sodium fluoride, barium fluosilicate, sodium silicofluoride, and cryolite. Cryolite has returned in recent years as a relatively safe fruit and vegetable insecticide, used in integrated pest management Programmes. The fluoride ion inhibits many enzymes that contain iron, calcium, and magnesium. Several of these enzymes are involved in energy production in cells, as in the case of phosphatases and phosphorylases.

Boric acid, used against cockroaches and other crawling household pests in the 1930's and '40's, has also returned. As a salt, it is non-volatile and will remain effective as long as it is kept dry and in adequate concentration. Consequently, it has the longest residual activity of any insecticide used for crawling household insects, and is quite useful in the control of all cockroach species when placed in wall voids and other protected, difficult-to-reach sites. It acts as a stomach poison and insect cuticle wax absorber.

Sodium borate (disodium octaborate tetrahydrate) resembles boric acid in its action. This water-soluble salt is used to treat lumber and other wood products to control decay fungi, termites, and other wood infesting pests. The last group of inorganics is the silica gels or silica aerogels–light, white, fluffy, silicate dusts used for household insect control. The silica aerogels kill insects by absorbing waxes from the insect cuticle, permitting the continuous loss of water from the insect body, causing the insects to become desiccated and die from dehydration. These include Dri-Die, Drianone, and Silikil Microcel. Drianone is fortified with pyrethrum and synergists to enhance its effectiveness.

NEW-MISC. INSECTICIDE CLASSES

Seven classes of insecticides have made their appearance in recent years. These are summarized below.

METHOXYACRYLATES

Fluacrypyrin is an acaricide for fruit and is the only example of this class currently. It is registered for use on fruit in Japan.

NAPHTHOQUINONES

Acequinocyl is a miticide with insecticidal activity for pome fruit, nut crops, citrus and ornamentals. It is the only member of this group at present and the mode of action is not yet determined. It holds registrations in Korea and Japan but not in the US.

ACEQUINOCYL (Kanemite, Piton)

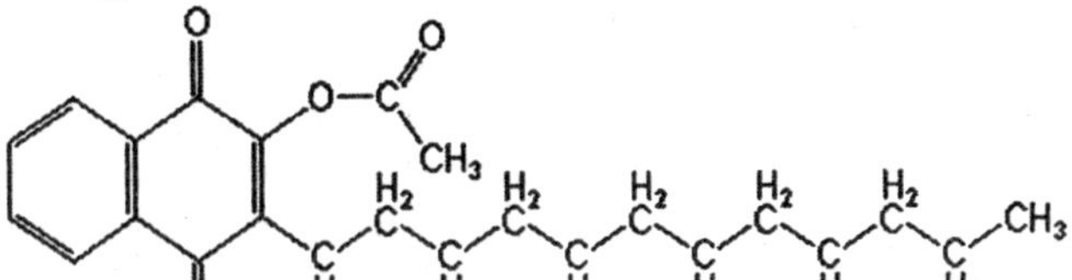

3-dodecyl-1,4-dihydro-1,4-dioxo-2-naphthyl acetate

FLUACRYPYRIM

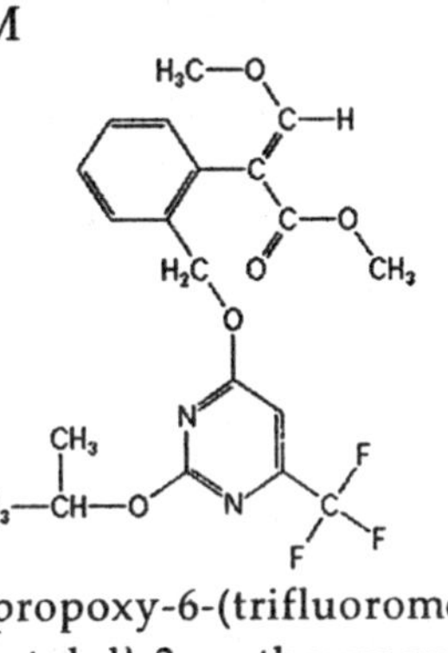

Methyl(*E*)-2-{a-[2-isopropoxy-6-(trifluoromethyl)pyrimidin-4-yloxy]-o-tolyl}-3-methoxyacrylate

NEREISTOXIN ANALOGUES

It includes thiocyclam, cartap, bensultap, and thiocytap-sodium. Analogues of nereistoxin have been known for decades. They generally are stomach poisons with some contact action and often show some systemic action. A major share of the development and use of these compounds has taken place in Japan. They are based on a natural toxin of the marine worm *Lumbriconereis heteropoda.* Of the many analogs synthesized only those that were metabolized back to the original nereistoxin after application were active.

In this sense members of this class are *proinsecticides* in that they are applied in their manufactured form but are known to degrade to a specific active component. The members of this group tend to be selectively active on Colopteran and Lepidopteran insect pests. Cartap is a broad spectrum insecticide with good activity against rice stem borer. Bensultap is used to control the Colourado Potato beetle and other insect pests. Thiosultap-sodium is used to control selected beetle and Lepidopteran pests on rice, vegetables and fruit trees.

Thiocyclam, is used for the control of similar pests in several crops. Members of this class act as acetyl choline receptor agonists at low concentrations and as channel blockers at higher concentrations. Although there has been commercial interest in thiocyclam for use in the US we do not believe there are commercial examples that are to achieve U.S. registration.

THIOCYCLAM (Evisect)

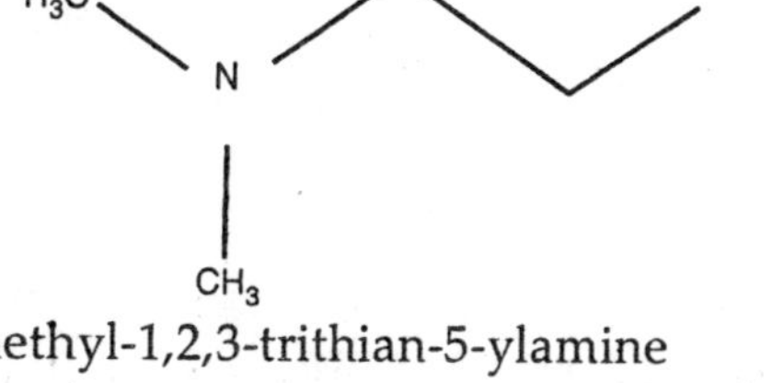

S S S H_3C N CH_3

N,N-dimethyl-1,2,3-trithian-5-ylamine

PYRIDINE AZOMETHINE

Pymetrozine, first registered in 1999 by EPA has a unique mode of action that is not fully understood. It appears to act by preventing insects from the Order Homoptera from inserting their stylus into plant tissue. Pymetrozine is used to control aphids and whiteflies in vegetables, potatoes, tobacco, deciduous citrus fruit hops and ornamentals.

PYMETROZINE (Fulfill)

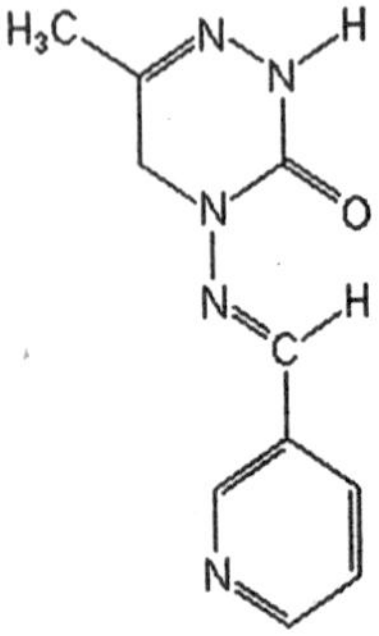

(*E*)-4,5-dihydro-6-methyl-4-(3-pyridylmethyleneamino)-1,2,4-triazin-3(2*H*)-one

PYRIMIDINAMINES

Pyrimidifen (Miteclean) is an insecticide and miticide. As a miticide the product controls spider and rust mites in deciduous fruits, citrus, vegetables and tea. As an insecticide it controls diamondback moth in vegetables. Very little information is available on the other member of this class, Flufenerim (S-1560), other than it is insecticidal.

PYRIMIDIFEN (Miteclean)

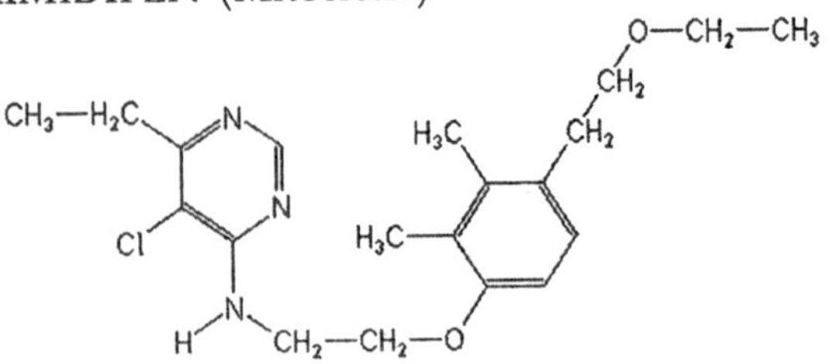

5-chloro-*N*-{2-[4-(2-ethoxyethyl)-2,3 dimethylphenoxy]ethyl} -6-ethylpyrimidin-4-amine

TETRONIC ACIDS

Spirodiclofen (Envidor) and spiromesifen (Oberon) are the only two members of this recently introduced class. Spirodiclofen has broad-spectrum activity against mites, and controls scale crawlers and psyllad nymphs. Action is good on eggs and quiescent stages. Target crops are citrus, grapes, nuts, pome and stone fruits.

SPIRODICLOFEN (Envidor)

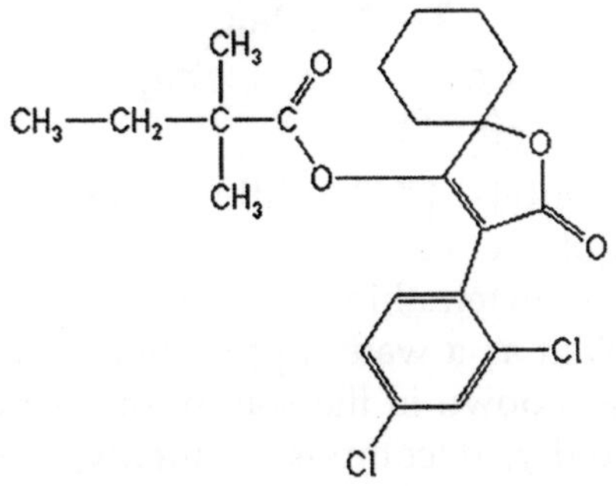

3-(2,4-dichlorophenyl)-2-oxo-1-oxaspiro dec-3-en-4-yl 2,2-dimethylbutyrate

MISCELLANEOUS COMPOUNDS

Clofentezine (Apollo, Acaristop), belongs to the unique group, the tetrazines, used as an acaricide/ovicide for deciduous fruits, citrus, cotton, cucurbits, vines and ornamentals. A newer and somewhat similar agent is etoxazole (TerraSan) which is an acaricide registered in 2002 for ornamentals grown in greenhouses. The modes of action of these two compounds are not yet understood.

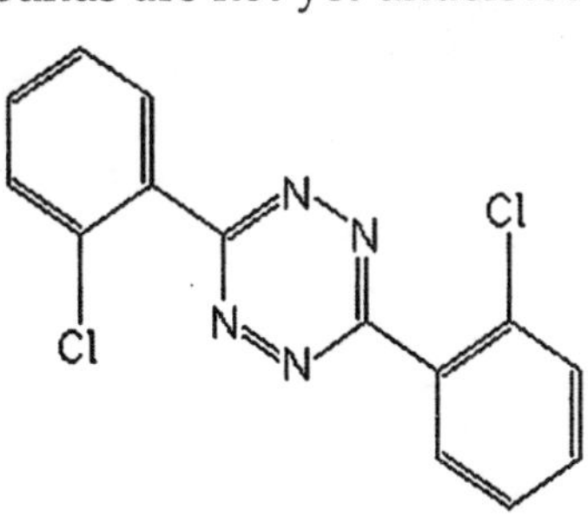

ETOXAZOLE (TerraSan)

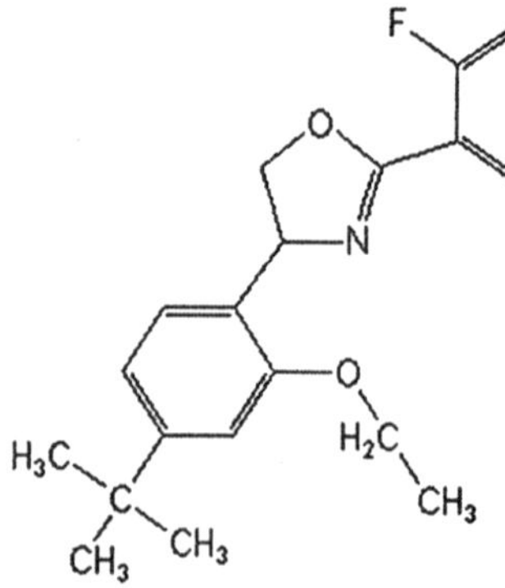

(*RS*)-5-tert-butyl-2-[2-(2,6-difluorophenyl)-4,5-dihydro-1,3-oxazol-4-yl]phenetole

Enzone, sodium tetrathiocarbonate, is used only on grapes and citrus applied as a water application and irrigated into the soil. It breaks down in the soil to form carbon disulfide, which acts rapidly, decomposes quickly, and is effective against nematodes, soil insects, and soil borne diseases.

PYRIDALYL (S-1812)

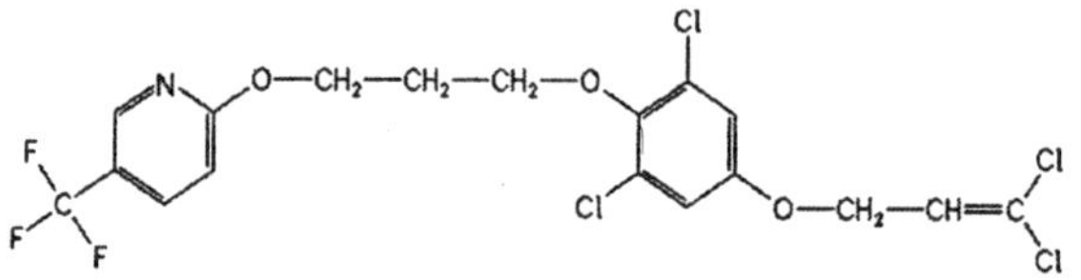

2,6-dichloro-4-(3,3-dichloroallyloxy)phenyl 3-[5-(trifluoromethyl)-2-pyridyloxy]propyl ether

AMIDOFLUMET (S-1955)

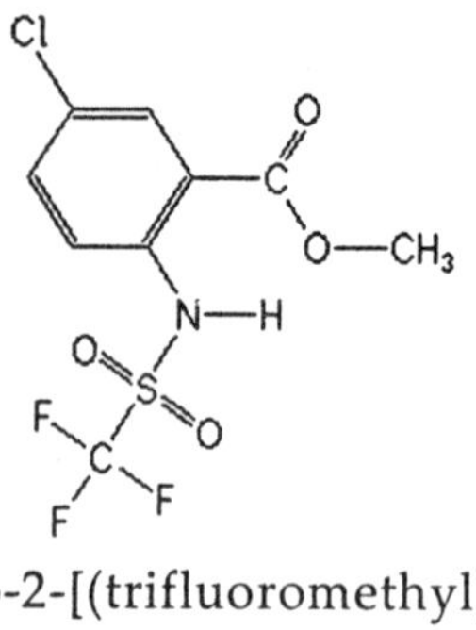

Methyl5-chloro-2-[(trifluoromethyl) sulfonyl] amino}

benzoate. The newest agents in this category are pyridanyl and amidoflumet. Pryidalyl (S-1812) is active on Lepidoptera and thrips and has the advantage of being active against pyrethroid-resistant insects. Little more is available on amidoflumet (S-1955) other than it is an acaricide in the early stage of its development.

BIORATIONAL INSECTICIDES

The U.S. EPA identifies biorational pesticides as inherently different from conventional pesticides, having fundamentally different modes of action, and consequently, lower risks of adverse effects from their use. Biorational has come to mean any substance of natural origin (or man-made substances resembling those of natural origin), that has a detrimental or lethal effect on specific target pest(s), e.g., insects, weeds, plant diseases (including nematodes), and vertebrate pests, possess a unique mode of action, are non-toxic to man and his domestic plants and animals, and have little or no adverse effects on wildlife and the environment. EPA uses a similar term, biopesticides, which will be defined below.

Biorational insecticides are grouped as either

- Biochemicals (hormones, enzymes, pheromones and natural agents, such as insect and plant growth regulators),
- Microbial (viruses, bacteria, fungi, protozoa, and nematodes).

In the 1990's the US-EPA began to emphasize a class of products known as biopesticides. EPA places biopesticides into three categories

- Microbial pesticides (bacteria, fungi, virus or protozoa)
- Biochemicals – natural substances that control pests by non-toxic mechanisms. An example is insect pheromones.
- Plant-Incorporated protectants (PIPs) – (primarily transgenic plants, e.g., Bt corn).

EPA discloses that at the end of 2001 there were nearly 200 biopesticide active ingredients registered comprising nearly 800 products.

Characteristics that distinguish biorational and biopesticides from conventional ones include very low orders of toxicity to non-target species, pest targets are specific, generally low use rates, rapid decomposition in the environment, usually work well in IPM Programmes and reduce reliance on conventional pesticide products.

The terms "biorational" and "biopesticide" overlap but are not identical. Below is an overview of what is considered as biorational insecticides. In some cases there are overlaps with botanicals (e.g., rotenone, florals, etc. and also conventional insecticides (e.g., benzoylureas). We will point out the discrepancies in classification between the biorational and biopesticide categories where they occur.

INSECT PHEROMONES

Most insects appear to communicate by releasing molecular quantities of highly specific compounds that vapourize readily and are detected by insects of the same species. These delicate molecules are known as pheromones. The word pheromone comes from the Greek pherein, "to carry," and hormon, "to excite or stimulate."

Of 1,314 species of insects with confirmed attraction responses to identified pheromones, 1,260 of these pheromones are produced by females. Only 54 species use male-produced sex attractants. In a few species both sexes produce the same attractant by both sexes.

Pheromones are classified as either, releasers and or primers. Releasers are fast-acting and are used by insects for sexual attraction, aggregation (including trail following), dispersion, oviposition, and alarm. Primers are slow-acting and cause gradual changes in growth and development, especially in social insects by regulating caste ratios of the colony.

The five principal uses for sex pheromones are

- Male trapping, to reduce the reproductive potential of an insect population;
- Movement studies, to determine how far and where insects move from a given point;

- Population monitoring, to determine when peak emergence or appearance occurs;
- Detection Programmes, to determine if a pest occurs in a limited trapping area, such as around international airports or quarantined areas;
- The "confusion or mating disruption" technique.

The first use of mating disruption involved gossyplure, the pink bollworm pheromone. Incorporated into small, hollow, polyvinyl fibres that permit slow release of the pheromone, it was broadcast heavily and uniformly over infested cotton fields. In mid 2002, EPA had registered 36 pheromones which comprised over 200 individual products.

Despite praise for the potential of sex pheromones, they are most practically used in survey traps to provide information about population levels, to delineate infestations, to monitor control or eradication Programmes, and to warn of new pest introductions.

INSECT GROWTH REGULATORS

Insect growth regulators (IGRs) are chemical compounds that alter growth and development in insects. The IGRs disrupt insect growth and development in three ways As juvenile hormones, as precocenes, and as chitin synthesis inhibitors. Juvenile hormones (JH) include ecdysone (the molting hormone), JH mimic, JH analog (JHA), and are known by their broader synonyms, juvenoids and juvegens. They disrupt immature development and emergence as adults. Precocenes interfere with the normal function of glands that produce juvenile hormones. And, chitin synthesis inhibitors, (conventional benzoylureas, buprofezin and cyromazine), affect the ability of insects to produce new exoskeletons when molting.

The IGRs are effective when applied in very minute quantities and generally have few or no effects on humans and wildlife. They are, however, nonspecific, since they affect not only the target species, other arthropods as well.

Instead of killing directly, IGRs interfere in the normal mechanisms of development and cause the insects to die before reaching the adult stage. One JH is the classical juvabione, found in the wood of balsam fir. Its effect was discovered quite by accident when paper towels made from this source were used to line insect-rearing containers, and the insects' development was suppressed.

Some of these plant-derived substances actually serve to inhibit the development of insects feeding thereon, thus protecting the host plant. These are referred to broadly as antijuvenile hormones, more accurately, antiallatotropins, or precocenes. Although the mode of action of the precocenes is still unclear, it is known that they depress the level of juvenile hormone below that normally found in immature insects.

For practical purposes, IGRs are used on crops to suppress damaging insect numbers. They would be applied with the purpose of preventing pupal development or adult emergence, thus keeping the insects in the immature stages, resulting eventually in their deaths. Commercial successful pheromones have shown activity on mosquito larvae, caterpillars, and hemipterans (bugs), although effects have been observed on practically all insect orders.

Methoprene (Altosid)

- Methylethyl (2E,4E)-11-methoxy-3,7,11-trimethyl-2,4-dodecadienoate

Fenoxycarb (Logic, Award, Comply, Torus) is a carbamate stomach insecticide that has also JH-type effects when contacted or ingested by a wide array of arthropod pests, e.g., ants, roaches, ticks, chiggers and many others. Pyriproxifen (Knack, Esteem, Admiral, Archer) is an effective molt inhibitor for a wide range of insects, but particularly useful for whitefly on cotton, citrus scales, fly-breeding sites such as livestock and poultry houses, and aquatic sites for mosquito control.

Another is buprofezin (Applaud) classed as a thiadiazine IGR. Both have given excellent results in controlling the whitefly complex, now a universal problem in U. S. cotton production.

PYRIPROXYFEN (Knack, Esteem)

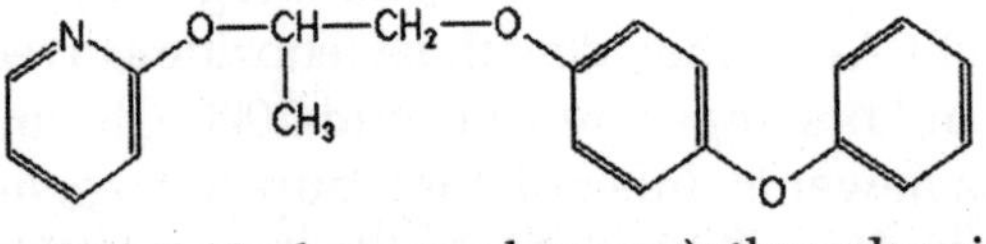

2-[1-methyl-2-(4-phenoxyphenoxy)ethoxy]pyridine

BUPROFEZIN (Applaud)

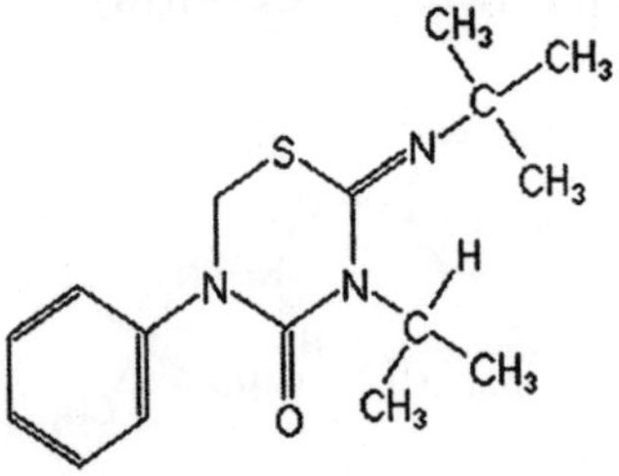

2-[(1,1-dimethylethyl)imino]tetrahydro-3-(1-methylethyl)-5-phenyl-4*H*-1,3,5-thiadiazin-4-one

None of the above– pyriproxifen, buprofezin, fenoxycarb or the methoprene, hydroprene group of juvenile hormone mimics are considered to be biopesticides by EPA.

HYDRAZINE INSECTICIDE/IGRS

A newer class of insecticidal IGRs is the hydrazines, which includes tebufenozide, halofenozide, methoxyfenozide and chromafenozide. All are ecdysone agonists or disruptors. EPA has not classified members of this group as biopesticides. Tebufenozide, in addition to being both a stomach and contact insecticide, has also JH-IGR characteristics. It disrupts the molting process by antagonizing ecdysone, the molting hormone. Lepidopteran pests are controlled while maintaining natural populations of beneficial insect predators and parasites. Halofenozide registered in1999, is a systemic IGR, effective on cutworms, sod webworms, armyworms and white grubs, and has some ovicidal activity.

It lacks the stomach or contact characteristics of tebufenozide.

Methoxyfenozide, like tebufenozide, is both a stomach and contact insecticide with JH-IGR qualities. It is systemic only through the roots. Pests controlled are lepidopterans such as codling moth, oriental fruit moth, European corn borer, and others. Crop candidates are cotton, corn, vegetables, pome fruit, and grapes. EPA considered methoxyfenozide as a reduced-risk candidate and first registered it in mid-2000. Chromafenozide (Matric) is a newer member of this group, not registered in the U.S., and is used to control various lepidopteran pests in vegetables and ornamentals.

TEBUFENOZIDE (Mimic, Confirm)

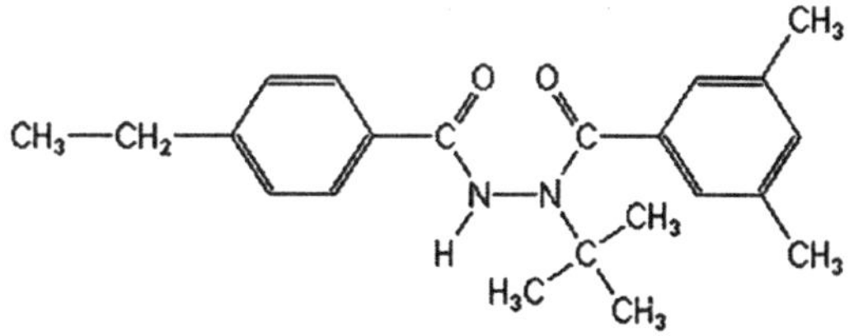

3,5-dimethylbenzoic acid 1-(1,1-dimethylethyl)-2-(4-ethylbenzoyl)hydrazide

OTHER BIORATIONAL INSECTICIDES

A number of the products that we covered under botanicals and florals are also considered by many to be biorational products, and indeed, EPA includes them under the biopesticide category. Some examples include Neem oil, cinnamaldehyde, and eugenol. A new product, Virtuoso, is a Streptomycetes-based agent that controls caterpillars but little is yet published on it, at present. Clandosan is a naturally occurring product derived from crab and shrimp shells and used as a nematicide. It is a dried, powdered, chitin protein isolated from crustacean exoskeletons and blended with urea. It stimulates growth of beneficial soil microorganisms that control nematodes, but does not have a direct adverse effect on nematodes as such.

MICROBIALS

Microbial insecticides obtain their name from

microorganisms that are used to control certain insects. The insect disease-causing microorganisms do not harm other animals or plants. At present there are relatively few produced commercially and approved by the EPA (over 55 natural, and 16 bioengineered organisms) for use on food and feed crops. In mid-2002, the EPA list of registered microbials included 35 bacteria, 1 yeast, 17 fungi, 1 protozoan, 6 viruses, 8 bioengineered organisms and 8 transgenic crop genes.

The insecticidal bacterium *Bacillus thuringiensis* (*Bt*) was discovered in the early 20th century. It occurs as a large number of subspecies that are identified among other characteristics by surface antigens, plasmid arrays, and breadth of species responding to its insecticidal action. *Bt* is a soil inhabiting, gram-positive sporulating bacterium that produces one or more very tiny parasporal crystals within its sporulating cells. These crystals are composed of large proteins known as delta-endotoxins. Delta-endotoxins act by binding to specific receptor sites on the gut epithelium, leading slowly to degradation of the gut lining and starvation. Thus, several days are required to kill insects that have ingested *Bt* products.

Over time, several *B. thuringiensis* varieties have been discovered, each with its distinct toxicity characteristics to different insect species. *B. thuringiensis* var. *kurstaki* was the first, being the spores and crystalline delta-endotoxin as the active ingredient, and produced by *B. thuringiensis* Berliner, var. *kurstaki*, Serotype H-3a3b, HD-1, in fermentation. Products from this process control most lepidopteran pests, the caterpillars with high gut pH, which include the armyworms, cabbage looper, imported cabbage worm, gypsy moth, and spruce budworm. The next was *B. thuringiensis* var. *israelensis*, being the crystalline delta-endotoxin as the active ingredient, and produced by fermentation of *B. thuringiensis* Berliner, var. *israelensis*, Serotype H-14. These products are used primarily for the control of aquatic insects, the mosquitoes and black flies in their larval forms.

Then came *B. thuringiensis var. aizawai*, produced by this variety, Serotype H-7, in fermentation. This product is currently registered only for the control of the wax moth larval infestations in the honey comb of honey bees. Following this came *B. thuringiensis* var. *morrisoni*, spores and delta-endotoxin produced by fermentation of Serotype 8a8b.

This is again a broad spectrum Bt for most caterpillars on most crops including the home garden. *B. thuringiensis* var. *san diego* was developed for Colourado potato beetle control on all its hosts, the elm leaf beetle and other beetle larvae on a wide range of shade and ornamental trees.

This was the first Bt product that was effective against coleopteran larvae. *B. thuringiensis* var. *tenebrionis* and the identical var. *san diego* were also developed for the Colourado potato beetle.The utilization of Bt genes transplanted into crops, which is addressed elsewhere in this document, is transforming the area of microbial pesticides

An innovative development in the agricultural use of microbial insecticides was the addition of feeding or gustatory stimulants, making the mixtures serves as baits. The feeding stimulants attracted the caterpillars to treated foliage, which increased their consumption of the microbial.

FUNGI

Mycar was a promising biorational miticide, a mycoacaricide, but was discontinued by the manufacturer in 1984. The microorganism was *Hirsutella thompsonii*, a parasitic fungus that infects and kills the citrus rust mite.

Under optimum conditions *H. thompsonii* can infect spider mites and other nontarget mites. It was, however, consistently effective only against the citrus rust mite; thus a selective miticide.

EPA registered *Metarhizium anisopliae* St. F52 in mid-2002 to control various ticks, beetles, flies, gnats and thrips for non-food outdoor and greenhouse uses. Certain of the registered uses were conditional for two years pending results of tick performance studies.

Another strain of this organism (St. ESF1) is also registered as a termiticide. Application was made in 1998 to register the fungus *Aspergillus flavus* strain AF36 as a bioinsecticide for cotton. Its purpose is to help reduce the incidence of other

Aspergillus spp. that produce the highly toxic mycotoxin, aflatoxin, in cotton seed.

PROTOZOA

Nosema locustae is a biorational originally developed by Sandoz, Inc., in 1981, for the control of grasshoppers. Marketed under the names, and Grasshopper Attack, the microorganism is a protozoan. These have been discontinued although the registrations remain.

NEMATODES

There were two commercial nematode products available for termite control. The nematode, *Neoaplectana carpocapsae,* in the family Steirnernamatidae, is specific for subterranean termites. It kills all stages of these termites by delivering a pathogenic bacterium, *Xenorhabdus* spp., which is lethal within 48 hours after penetration. Unfortunately, neither product succeeded commercially.

A NEW HORIZON TRANSGENIC PESTICIDES

The concept that organisms and crops could be engineered to augment pest control was well known in the 1970's. By the mid 1980s, one company, Monsanto had committed to a research Programme designed to create crop protection products through the application of biotechnology. Charles has produced a very readable history of pesticide-related transgenic crops and this book is recommended to those who want to understand how this new technology unfolded.

Transgenic organisms are genetically altered by artificial introduction of DNA from another organism. The artificial gene sequence is referred to as a *transgene.* Plants with such transgenes are also referred to as being *genetically modified* (GM). Plants that emulate insecticides are those altered to induce *insect-resistance* (also called *plant pesticides* or *plant incorporated protectants.* The purpose of the following paragraphs is to summarize what biotechnology has contributed to insecticide science in the course of just the last decade or so.

Research built on the elucidation of the genetic code in the early 1950's and culminating in the 1990's allowed those using the techniques of biotechnology to move genes coding for specific traits from selected organism to crop cells. Such altered cells were then regenerated to viable crop plants through tissue culture. Several transgenic crops have thus been and are being created from backcrossing the selected traits into elite seed lines. The result has led us to *plant pesticides*.

Plant pesticides are defined by EPA as plants that have been genetically engineered to contain the delta-endotoxin genes from *Bacillus thuringiensis*. This definition will expand as genes from additional sources are incorporated into plants.

In 1995, EPA registered the first plant pesticide. It was Monsanto's Bt-cotton containing B. *t. Cry1Ac* delta-endotoxin, following more than a decade of research. This novel form of cotton was introduced experimentally in 1995 as Bollgard cotton, resistant to tobacco budworm, cotton bollworm, and pink bollworm with activity on other minor lepidopteran pests. *Bt*-enhanced cotton, corn and other insect resistant crops produce one or more crystalline proteins that disrupt the gut lining of susceptible insect pests feeding on their tissues which cause the pests to stop feeding and die. Several plant pesticides have been introduced in the U.S. since 1995. Some of these have been very successful commercially while others, such as, NewLeaf Potatoes, have been withdrawn from the market.

In some subsequent product introductions the performance of these plant pesticides have been enhanced or augmented by use of *stacked genes*. This means that more than one transgene is introduced into the same crop to achieve multiple desired characteristics.

In the U.S. the proportion of total planted acres of *B.t.* corn, grew from 18% in 2000 to 26 % in 2003, and this does not include stacked-gene plant insecticides. The proportions for cotton were even higher if stacked-gene varieties were included.

Three U.S. federal agencies (USDA, FDA and EPA) regulate the release and use of transgenic plants and plant pesticides under a coordinated framework. There is a sharing and partitioning of biotechnology regulatory responsibility among these three agencies. The approach of each agency is similar, in that, they each evaluate risks from a sound science perspective and regulate individual products on a case-by-case basis. There is considerable reliance on comparing transgenic organisms with their conventional counterparts that have a known history of safe use. The emphasis during the regulatory review is to assure that the GM organism will not produce harmful toxins, allergens, etc., and will not, once released, cause adverse effects (such as become pests themselves).

Chapter 7

Pesticides

A pesticide is a substance or mixture of substances used for preventing, controlling, or lessening the damage caused by a pest. A pesticide may be a chemical substance, biological agent (such as a virus or bacteria), antimicrobial, disinfectant or device used against any pest. Pests include insects, plant pathogens, weeds, molluscs, birds, mammals, fish, nematodes (roundworms) and microbes that compete with humans for food, destroy property, spread or are a vector for disease or cause a nuisance.

Although there are benefits to the use of pesticides, there are also drawbacks, such as potential toxicity to humans and other animals.Pesticides are hazadous to some wildlife in the sea because it gets evapourated and goes into the clouds.Then it rains, surface run-off into the sea and poisions them.

TYPES OF PESTICIDES

There are multiple ways of classifying pesticides.

- Algicides or Algaecides for the control of algae
- Avicides for the control of birds
- Bactericides for the control of bacteria
- Fungicides for the control of fungi and oomycetes
- Herbicides for the control of weeds
- Insecticides for the control of insects - these can be Ovicides (substances that kill eggs), Larvicides (substances that kill larvae) or Adulticides (substances that kill adult insects)
- Miticides or Acaricides for the control of mites
- Molluscicides for the control of slugs and snails

- Nematicides for the control of nematodes
- Rodenticides for the control of rodents
- Virucides for the control of viruses (e.g. H5N1)

Pesticides can also be classed as synthetic pesticides or biological pesticides (biopesticides), although the distinction can sometimes blur.

Broad-spectrum pesticides are those that kill an array of species, while narrow-spectrum, or selective pesticides only kill a small group of species.

A systemic pesticide moves inside a plant following absorption by the plant. With insecticides and most fungicides, this movement is usually upward (through the xylem) and outward. Increased efficiency may be a result. Systemic insecticides which poison pollen and nectar in the flowers may kill needed pollinators such as bees.

Most pesticides work by poisoning pests.

USES, BENEFITS AND DRAWBACKS

Pesticides are used to control organisms which are considered harmful.For example, they·are used to kill mosquitoes that can transmit potentially deadly diseases like west nile virus and malaria. They can also kill bees, wasps or ants that can cause allergic reactions. Insecticides can protect animals from illnesses that can be caused by parasites such as fleas.Pesticides can prevent sickness in humans that could be caused by mouldy food or diseased produce. Herbicides can be used to clear roadside weeds, trees and brush.

They can also kill invasive weeds in parks and wilderness areas which may cause environmental damage. Herbicides are commonly applied in ponds and lakes to control algae and plants such as water grasses that can interfere with activities like swimmng and fishing and cause the water to look or smell unpleasant. Uncontrolled pests such as termites and mould can damage structures such as houses.

Pesticides are used in grocery stores and food storage facilities to manage rodents and insects that infest food such as grain. Each use of a pesticide carries some associated risk. Proper pesticide use decreases these associated risks to a level deemed acceptable by pesticide regulatory agencies such as the EPA and PMRA. Pesticides can save farmers money by preventing crop losses to insects and other pests; in the US, farmers get an estimated four-fold return on money they spend on pesticides. One study found that not using pesticides reduced crop yields by about 10%.

DDT, sprayed on the walls of houses, has been used to fight malaria since the 1950s. Recent policy statements by the World Health Organization have given stronger support to this approach. Dr. Arata Kochi, WHO's malaria chief, said, "One of the best tools we have against malaria is indoor residual house spraying. Of the dozen insecticides WHO has approved as safe for house spraying, the most effective is DDT."However, since then, an October 2007 study has linked breast cancer from exposure to DDT prior to puberty.

Scientists estimate that DDT and other chemicals in the organophosphate class of pesticides have saved 7 million human lives since 1945 by preventing the transmission of diseases such as malaria, bubonic plague, sleeping sickness, and typhus. However, DDT use is not always effective, as resistance to DDT was identified in Africa as early as 1955, and by 1972 nineteen species of mosquito worldwide were resistant to DDT. A study for the World Health Organization in 2000 from Vietnam established that non-DDT malaria controls were significantly more effective than DDT use.

In the US, about a quarter of pesticides used are used in houses, yards, parks, golf courses, and swimming pools. About 70% of the pesticides sold in the US are used in agriculture.

Since before 2500 BC, humans have utilized pesticides to protect their crops. The first known pesticide was elemental sulfur dusting used in Sumeria about 4,500 years ago. By the 15th century, toxic chemicals such as arsenic, mercury and lead were being applied to crops to kill pests. In the 17th century, nicotine sulfate was extracted from tobacco leaves for use as an insecticide. The 19th century saw the introduction of two more natural pesticides, pyrethrum which is derived from chrysanthemums, and rotenone which is derived from the roots of tropical vegetables.

In 1939, Paul Müller discovered that DDT was a very effective insecticide. It quickly became the most widely-used pesticide in the world.

In the 1940s manufacturers began to produce large amounts of synthetic pesticides and their use became widespread. Some sources consider the 1940s and 1950s to have been the start of the "pesticide era." Pesticide use has increased 50-fold since 1950 and 2.5 million tons (2.3 million metric tons) of industrial pesticides are now used each year. Seventy-five percent of all pesticides in the world are used in developed countries, but use in developing countries is increasing.

In the 1960s, it was discovered that DDT was preventing many fish-eating birds from reproducing, which was a serious threat to biodiversity. Rachel Carson wrote the best-selling book *Silent Spring* about biological magnification. DDT is now banned in at least 86 countries, but it is still used in some developing nations to prevent malaria and other tropical diseases by killing mosquitoes and other disease-carrying insects.

REGULATION

In most countries, in order to sell or use a pesticide, it must be approved by a government agency. For example, in the United States, the Environmental Protection Agency (EPA) does so. Complex and costly studies must be conducted to indicate whether the material is safe to use and effective against the intended pest. During the registration process, a label is created which contains directions for the proper use of the material. Based on acute toxicity, pesticides are assigned to a Toxicity Class.

Some pesticides are considered too hazardous for sale to the general public and are designated restricted use pesticides. Only certified applicators, who have passed an exam, may purchase or supervise the application of restricted use pesticides. Records of sales and use are required to be maintained and may be audited by government agencies charged with the enforcement of pesticide regulations.

In Canada, over 140 municipalities and the entire province of Quebec have now placed restrictions on the cosmetic use of synthetic lawn pesticides as a result of health and environmental concerns. The Ontario provincial government promised on September 24, 2007 to also implement a province-wide ban on the cosmetic use of lawn pesticides, for protecting the public. Medical and environmental groups support such a ban. On April 22, 2008, the Provincial Government of Ontario announced that it will pass legislation that will prohibit, province-wide, the cosmetic use and sale of lawn and garden pesticides.

The Ontario legislation would also echo Massachusetts law requiring pesticide manufacturers to reduce the toxins they use in production. The Province of Prince Edward Island is also considering such legislation. On April 3, 2008, the Canadian Cancer Society released opinion poll results conducted by Ipsos Reid, which established that a clear majority of residents in the provinces of British Columbia and Saskatchewan want province-wide cosmetic lawn pesticide bans, and that the majority of respondents believe that cosmetic pesticides are a threat to their health.

Though pesticide regulations differ from country to country, pesticides and products on which they were used are traded across international borders. To deal with inconsistencies in regulations among countries, delegates to a conference of the United Nations Food and Agriculture Organization adopted an International Code of Conduct on the Distribution and Use of Pesticides in 1985 to create voluntary standards of pesticide regulation for different countries. The Code was updated in 1998 and 2002. The FAO claims that the code has raised awareness about pesticide hazards and decreased the number of countries without restrictions on pesticide use.

Two other efforts to improve regulation of international pesticide trade are the United Nations London Guidelines for the Exchange of Information on Chemicals in International Trade and the United Nations Codex Alimentarius Commission. The former seeks to implement procedures for ensuring that prior informed consent exists between countries buying and selling pesticides, while the latter seeks to create uniform standards for maximum levels of pesticide residues among participating countries. Both initiatives operate on a voluntary basis.

Reading and following label directions is required by law in countries such as the US and in limited parts of the rest of the world.

In the US, the Federal Insecticide, Fungicide, and Rodenticide Act (FIFRA) was first passed in 1947, giving the United States Department of Agriculture responsibility for regulating pesticides. In 1972, FIFRA underwent a major revision and transferred responsibility of pesticide regulation to the Environmental Protection Agency and shifted emphasis to protection of the environment and public health.

One study found pesticide self-poisoning the method of choice in one third of suicides worldwide, and recommended, among other things, more restrictions on the types of pesticides that are most harmful to humans.

ENVIRONMENTAL EFFECTS

Pesticide use raises a number of environmental concerns. Over 98% of sprayed insecticides and 95% of herbicides reach a destination other than their target species, including non-target species, air, water, bottom sediments, and food. Pesticide drift occurs when pesticides suspended in the air as particles are carried by wind to other areas, potentially contaminating them. Pesticides are one of the causes of water pollution, and some pesticides are persistent organic pollutants and contribute to soil contamination.

HEALTH EFFECTS

Pesticides can present danger to consumers, bystanders, or workers during manufacture, transport, or during and after use. The American Medical Association recommends limiting exposure to pesticides and using safer alternatives

Particular uncertainty exists regarding the long-term effects of low-dose pesticide exposures. Current surveillance systems are inadequate to characterize potential exposure problems related either to pesticide usage or pesticide-related illnesses...Considering these data gaps, it is prudent...to limit pesticide exposures...and to use the least toxic chemical pesticide or non-chemical alternative.

FARMERS AND WORKERS

There have been many studies of farmers with the goal of determining the health effects of pesticide exposure. The World Health Organisation and the UN Environment Programme estimate that each year, 3 million workers in agriculture in the developing world experience severe poisoning from pesticides, about 18,000 of whom die. According to one study, as many as 25 million workers in developing countries may suffer mild pesticide poisoning yearly.

Organophosphate pesticides have increased in use, because they are less damaging to the environment and they are less persistent than organochlorine pesticides. These are associated with acute health problems for workers that handle the chemicals, such as abdominal pain, dizziness, headaches, nausea, vomiting, as well as skin and eye problems. Additionally, many studies have indicated that pesticide exposure is associated with long-term health problems such as respiratory problems, memory disorders, dermatologic conditions, cancer, depression, neurological deficits, miscarriages, and birth defects. Summaries of peer-reviewed research have examined the link between pesticide exposure and neurologic outcomes and cancer, perhaps the two most significant things resulting in organophosphate-exposed workers.

There are concerns that pesticides used to control pests on food crops are dangerous to people who consume those foods. These concerns are one reason for the organic food movement. Many food crops, including fruits and vegetables, contain pesticide residues after being washed or peeled. Chemicals that are no longer used but which are resistant to breakdown for long periods may remain in soil and water and thus in food. The United Nations Codex Alimentarius

Commission has recommended international standards for Maximum Residue Limits (MRLs), for individual pesticides in food.

In the EU, MRLs are set by DG-SANCO. In the US, levels of residues that remain on foods are limited to tolerance levels that are established by the US EPA and are considered safe. The EPA sets the tolerances based on the toxicity of the pesticide and its breakdown products, the amount and frequency of pesticide application, and how much of the pesticide (i.e., the residue) remains in or on food by the time it is marketed and prepared.

Tolerance levels are obtained using scientific risk assessments that pesticide manufacturers are required to produce by conducting toxicological studies, exposure modeling and residue studies before a particular pesticide can be registered, however, the effects are tested for single pesticides, and there is little information on possible synergistic effects of exposure to multiple pesticide traces in the air, food and water.

A study published by the United States National Research Council in 1993 determined that for infants and children, the major source of exposure to pesticides is through diet. A study in 2006 measured the levels of organophosphorus pesticide exposure in 23 school children before and after replacing their diet with organic food (food grown without synthetic pesticides). In this study it was found that levels of organophosphorus pesticide exposure dropped dramatically and immediately when the children switched to an organic diet.

In the US, the National Academy of Sciences estimates that between 4,000 and 20,000 cases of cancer are caused per year by pesticide residues in food in allowable amounts.

The Pesticide Data Programme, a Programme started by the United States Department of Agriculture is the largest tester of pesticide residues on food sold in the United States. It began in 1991, and has since tested over 60 different types of food for over 400 different types of pesticides - with samples collected close to the point of consumption. Their most recent summary results are from the year 2005

To reduce the amounts of pesticide residues in food, consumers can wash, peel, and cook their food; trim the fat from meat; and eat a variety of foods to avoid repeat exposure to a pesticide typically used on a given crop. Consumers can also buy food that is grown organically, though even organic food may have traces of pesticides. Exposure routes other than consuming food that contains residues, in particular pesticide drift, are potentially significant to the general public.

The Bhopal disaster occurred when a pesticide plant released 40 tons of methyl isocyanate (MIC) gas, intermediate chemical in the production of some pesticides. The disaster immediately killed nearly 3,000 people and ultimately caused at least 15,000 deaths. In China, an estimated half million people are poisoned by pesticides each year, 500 of whom die.

Children have been found to be especially susceptible to the harmful effects of pesticides. A number of research studies have found higher instances of brain cancer, leukemia and birth defects in children with early exposure to pesticides, according to the Natural Resources Defense Council.

Peer-reviewed studies now suggest neurotoxic effects on developing animals from organophosphate pesticides at legally-tolerable levels, including fewer nerve cells, lower birth weights, and lower cognitive scores.The EPA finished a 10 year review of the organophosphate pesticides following the 1996 Food Quality Protection Act, but did little to account for developmental neurotoxic effects, drawing strong criticism from within the agency and from outside researchers.

Some scientists think that exposure to pesticides in the uterus may have negative effects on a fetus that may manifest as problems such as growth and behavioural disorders or reduced resistance to pesticide toxicity later in life.

A new study conducted by the Harvard School of Public Health in Boston, has discovered a 70% increase in the risk of developing Parkinson's disease for people exposed to even low levels of pesticides.

A 2008 study from Duke University found that the Parkinson's patients were 61 percent more likely to report direct pesticide application than were healthy relatives. Both insecticides and herbicides significantly increased the risk of Parkinson's disease.One study found that use of pesticides may be behind the finding that the rate of birth defects such as missing or very small eyes is twice as high in rural areas as in urban areas. Another study found no connection between eye abnormalities and pesticides. Pyrethrins, insecticides commonly used in common bug killers, can cause a potentially deadly condition if breathed in.

Pesticide safety education and pesticide applicator regulation are designed to protect the public from pesticide misuse, but do not eliminate all misuse. Reducing the use of pesticides and choosing less toxic pesticides may reduce risks placed on society and the environment from pesticide use. Integrated pest management, the use of multiple approaches to control pests, is becoming widespread and has been used with success in countries such as Indonesia, China, Bangladesh, the US, Australia, and Mexico.

IPM attempts to recognize the more widespread impacts of an action on an ecosystem, so that natural balances are not upset. New pesticides are being developed, including biological and botanical derivatives and alternatives that are thought to reduce health and environmental risks. In addition, applicators are being encouraged to consider alternative controls and adopt methods that reduce the use of chemical pesticides. Pesticides can be created that are targeted to a specific pest's life cycle, which can be more environmentally-friendly. For example, potato cyst nematodes emerge from their protective cysts in response to a chemical excreted by potatoes; they feed on the potatoes and damage the crop. A similar chemical can be applied to fields early, before the potatoes are planted, causing the nematodes to emerge early and starve in the absence of potatoes.

ALTERNATIVES

Alternatives to pesticides are available and include methods of cultivation, use of other organisms to kill pests, genetic engineering, and methods of interfering with insect breeding. Cultivation practices include polyculture (growing multiple types of plants), crop rotation, planting crops in areas where the pests that damage them do not live timing planting according to when pests will be least problematic, and use of trap crops that attract pests away from the real crop.

In the US, farmers have had success controlling insects by spraying with hot water at a cost that is about the same as pesticide spraying. Release of other organisms that fight the pest is another example of an alternative to pesticide use. These organisms can include natural predators or parasites of the pests. Biological pesticides based on entomopathogenic fungi, bacteria and viruses cause disease in the pest species can also be used. Interfering with insects' reproduction can be accomplished by sterilizing males of the target species and releasing them, so that they mate with females but do not produce offspring. This technique was first used on the screwworm fly in 1958 and has since been used with the medfly, the tsetse fly, and the gypsy moth. However, this can be a costly, time consuming approach that only works on some types of insects.

Some evidence shows that alternatives to pesticides can be equally effective as the use of chemicals. For example, Sweden has halved its use of pesticides with hardly any reduction in crops. In Indonesia, farmers have reduced pesticide use on rice fields by 65% and experienced a 15% crop increase.

HERBICIDE

A herbicide is used to kill unwanted plants. Selective herbicides kill specific targets while leaving the desired crop relatively unharmed. Some of these act by interfering with the growth of the weed and are often based on plant hormones. Herbicides used to clear waste ground are nonselective and kill all plant material with which they come into contact. Some plants produce natural herbicides, such as the genus Juglans (walnuts). They are applied in total vegetation control (TVC) Programmes for maintenance of highways and railroads. Smaller quantities are used in forestry, pasture systems, and management of areas set aside as wildlife habitat.

Herbicides are widely used in agriculture and in landscape turf management. In the U.S., they account for about 70% of all agricultural pesticide use. Prior to the widespread use of chemical herbicides, cultural controls, such as altering soil pH, salinity, or fertility levels, were used to control weeds. Mechanical control (including tillage) was also (and still is) used to control weeds.

The first widely used herbicide was 2,4-dichlorophenoxyacetic acid, often abbreviated 2,4-D. It was first commercialized by the Sherwin-Williams Paint company and saw use in the late 1940s. It is easy and inexpensive to manufacture, and kills many broadleaf plants while leaving grasses largely unaffected (although high doses of 2,4-D at crucial growth periods can harm grass crops such as maize or cereals). The low cost of 2,4-D has led to continued usage today and it remains one of the most commonly used herbicides in the world. Like other acid herbicides, current formulations utilize either an amine salt (usually trimethylamine) or one of many esters of the parent compound. These are easier to handle than the acid.

2,4-D exhibits relatively good *selectivity*, meaning, in this case, that it controls a wide number of broadleaf weeds while causing little to no injury to grass crops at normal use rates. A herbicide is termed selective if it affects only certain types of plants, and nonselective if it inhibits a very broad range of plant types. Other herbicides have been more recently developed that achieve higher levels of selectivity than 2,4-D.

The 1950s saw the introduction of the triazine family of herbicides, which includes atrazine, which have current distinction of being the herbicide family of greatest concern regarding groundwater contamination. Atrazine does not break down readily (within a few weeks) after being applied to soils of above neutral pH. Under alkaline soil conditions atrazine may be carried into the soil profile as far as the water table by soil water following rainfall causing the aforementioned contamination. Atrazine is said to have *carryover*, a generally undesirable property for herbicides.

Glyphosate, frequently sold under the brand name

Roundup, was introduced in 1974 for non-selective weed control. It is now a major herbicide in selective weed control in growing crop plants due to the development of crop plants that are resistant to it. The pairing of the herbicide with the resistant seed contributed to the consolidation of the seed and chemistry industry in the late 1990s.

Many modern chemical herbicides for agriculture are specifically formulated to decompose within a short period after application. This is desirable as it allows crops which may be affected by the herbicide to be grown on the land in future seasons. However, herbicides with low residual activity (i.e., that decompose quickly) often do not provide season-long weed control.

HEALTH EFFECTS

Certain herbicides affect metabolic pathways and systems unique to plants and not found in animals making many modern herbicides among the safest crop protection products having essentially no effect on mammals, birds, amphibians or reptiles. Some herbicides can cause a variety of health effects ranging from skin rashes to death. The pathway of attack can arise from intentional or unintentional direct consumption of the herbicide, improper application resulting in the herbicide coming into direct contact with people or wildlife, inhalation of aerial sprays, or food consumption prior to the labeled pre-harvest interval. Under extreme conditions herbicides can also be transported via surface runoff to contaminate distant water sources. Most herbicides decompose rapidly in soils via soil microbial decomposition, hydrolysis or photolysis and some herbicides are more persistent with longer soil half-lives.

Other alleged health effects can include chest pain, headaches, nausea and fatigue. All organic and non-organic herbicides must be extensively tested prior to approval for commercial sale and labeling by the Environmental Protection Agency. However, because of the large number of herbicides in use, there is significant concern regarding health effects. Some of the herbicides in use are known to be mutagenic, carcinogenic or teratogenic.

However, some herbicides may also have a therapeutic use. Current research aims to use herbicides as an anti-malaria drug that targets the plant-like apicoplast plastid in the malaria-causing parasite *Plasmodium falciparum*.

CLASSIFICATION OF HERBICIDES

Herbicides can be grouped by activity, use, chemical family, mode of action, or type of vegetation controlled.

By activity

- Contact herbicides destroy only the plant tissue in contact with the chemical. Generally, these are the fastest acting herbicides. They are less effective on perennial plants, which are able to regrow from rhizhomes, roots or tubers.
- Systemic herbicides are translocated through the plant, either from foliar application down to the roots, or from soil application up to the leaves. They are capable of controlling perennial plants and may be slower acting but ultimately more effective than contact herbicides.

By use

- Soil-applied herbicides are applied to the soil and are taken up by the roots of the target plant.
- Pre-plant incorporated herbicides are soil applied prior to planting and mechanically incorporated into the soil.
- Preemergent herbicides are applied to the soil before the crop emerges and prevent germination or early growth of weed seeds.
- Post-emergent herbicides are applied after the crop has emerged.

Their classification by mechanism of action (MOA) indicates the first enzyme, protein, or biochemical step affected in the plant following application. The main mechanisms of action are

- ACCase inhibitors are compounds that kill grasses. Acetyl coenzyme A carboxylase (ACCase) is part of the first step of lipid synthesis. Thus, ACCase inhibitors affect cell membrane production in the meristems of the grass plant. The ACCases of grasses are sensitive to these herbicides, whereas the ACCases of dicot plants are not.

- ALS inhibitors the acetolactate synthase (ALS) enzyme (also known as acetohydroxyacid synthase, or AHAS) is the first step in the synthesis of the branched-chain amino acids (valine, leucine, and isoleucine). These herbicides slowly starve affected plants of these amino acids which eventually leads to inhibition of DNA synthesis. They affect grasses and dicots alike. The ALS inhibitor family includes sulfonylureas (SUs), imidazolinones (IMIs), triazolopyrimidines (TPs), pyrimidinyl oxybenzoates (POBs), and sulfonylamino carbonyl triazolinones (SCTs). ALS is a biological pathway that exists only in plants and not in animals thus making the ALS-inhibitors among the safest herbicides.
- EPSPS inhibitors The enolpyruvylshikimate 3-phosphate synthase enzyme EPSPS is used in the synthesis of the amino acids tryptophan, phenylalanine and tyrosine. They affect grasses and dicots alike. Glyphosate (Roundup) is a systemic EPSPS inhibitor but inactivated by soil contact.
- Synthetic auxin inaugurated the era of organic herbicides. They were discovered in the 1940s after a long study of the plant growth regulator auxin. Synthetic auxins mimic this plant hormone. They have several points of action on the cell membrane, and are effective in the control of dicot plants. 2,4-D is a synthetic auxin herbicide.
- Photosystem II inhibitors reduce electron flow from water to NADPH2+ at the photochemical step in photosynthesis. They bind to the Qb site on the D1 protein, and prevent quinone from binding to this site. Therefore, this group of compounds cause electrons to accumulate on chlorophyll molecules. As a consequence, oxidation reactions in excess of those

normally tolerated by the cell occur, and the plant dies. The triazine herbicides (including atrazine) and urea derivatives (diuron) are photosystem II inhibitors.

ORGANIC HERBICIDES

Almost all herbicides in use today are considered "organic" herbicides in that they contain carbon as a primary molecular component. An notable exception would be the arsenical class of herbicides. Sometimes they are referred to as synthetic organic herbicides. Recently the term "organic" has come to imply products used in organic farming. Under this definition an organic herbicide is one that can be used in a farming enterprise that has been classified as organic. Organic herbicides are expensive and may not be affordable for commercial production. They are much less effective than synthetic herbicides and are generally used along with cultural and mechanical weed control practices.

Organic herbicides include

- Spices are now effectively used in patented herbicides.
- Vinegar is effective for 5-20% solutions of acetic acid with higher concentrations most effective but mainly destroys surface growth and so respraying to treat regrowth is needed. Resistant plants generally succumb when weakened by respraying.
- Steam has been applied commercially but is now considered uneconomic and inadequate. It kills surface growth but not underground growth and so respraying to treat regrowth of perennials is needed.
- Flame is considered more effective than steam but suffers from the same difficulties.

APPLICATION

Most herbicides are applied as water-based sprays using ground equipment. Ground equipment varies in design, but large areas can be sprayed using self-propelled sprayers equipped with a long boom, of 60 to 80 feet (20 to 25 m) with flat fan nozzles spaced about every 20 in (500 mm). Towed, handheld, and even horse-drawn sprayers are also used.

Synthetic organic herbicides can generally be applied aerially using helicopters or airplanes, and can be applied through irrigation systems (chemigation).

TERMINOLOGY

- Control is the destruction of unwanted weeds, or the damage of them to the point where they are no longer competitive with the crop.
- Suppression is incomplete control still providing some economic benefit, such as reduced competition with the crop.
- Crop Safety, for selective herbicides, is the relative absence of damage or stress to the crop. Most selective herbicides cause some visible stress to crop plants.

MAJOR HERBICIDES IN USE TODAY

- 2,4-D, a broadleaf herbicide in the phenoxy group used in turf and in no-till field crop production. Now mainly used in a blend with other herbicides that allow lower rates of herbicides to be used, it is the most widely used herbicide in the world, third most commonly used in the United States. It is an example of synthetic auxin(plant hormone).
- Atrazine, a triazine herbicide used in corn and sorghum for control of broadleaf weeds and grasses. Still used because of its low cost and because it works extrodinarily well on a broad spectrum of weeds common in the U.S. corn belt, Atrazine is commonly used with other herbicides to reduce the over-all rate of atrazine and to lower the for potential groundwater contamination, it is a photosystem II inhibitor.
- Clopyralid is a broadleaf herbicide in the pyridine group, used mainly in turf, rangeland, and for control of noxious thistles. Notorious for its ability to persist

in compost. It is another example of synthetic auxin.

- Dicamba, a persistent broadleaf herbicide active in the soil, used on turf and field corn. It is another example of synthetic auxin.
- Glufosinate ammonium, a broad-spectrum contact herbicide and is used to control weeds after the crop emerges or for total vegetation control on land not used for cultivation.
- Glyphosate, a systemic nonselective (it kills any type of plant) herbicide used in no-till burndown and for weed control in crops that are genetically modified to resist its effects. It is an example of an EPSPs inhibitor.
- Imazapyr, is a non-selective herbicide used for the control of a broad range of weeds including terrestrial annual and perennial grasses and broadleaved herbs, woody species, and riparian and emergent aquatic species.
- Imazapic, is a selective herbicide for both the pre- and post-emergent control of some annual and perennial grasses and some broadleaf weeds. Imazapic kills plants by inhibiting the production of branched chain amino acids (valine, leucine, and isoleucine), which are necessary for protein synthesis and cell growth.
- Linuron, is a non-selective herbicide used in the control of grasses and broadleafed weeds. It works by inhibiting photosynthesis.
- Metoalachlor, a pre-emergent herbicide widely used for control of annual grasses in corn and sorghum; it has largely replaced atrazine for these uses.
- Paraquat, a nonselective contact herbicide used for no-till burndown and in aerial destruction of marijuana and coca plantings. More acutely toxic to people than any other herbicide in widespread commercial use.
- picloram, a pyridine herbicide mainly used to control unwanted trees in pastures and edges of fields. It is another synthetic auxin.

- Triclopyr, a systemic, foliar herbicide in the pyridine group. It is used to control broadleaf weeds while leaving grasses and conifers unaffected.

HERBICIDES OF HISTORICAL INTEREST

- 2,4,5-Trichlorophenoxyacetic acid (2,4,5-T) was a widely used broadleaf herbicide until being phased out starting in the late 1970s. While 2,4,5-T itself is of only moderate toxicity, the manufacturing process for 2,4,5-T contaminates this chemical with trace amounts of 2,3,7,8-tetrachlorodibenzo-p-dioxin (TCDD). TCDD is extremely toxic to humans. With proper temperature control during production of 2,4,5-T, TCDD levels can be held to about.005 ppm. Before the TCDD risk was well understood, early production facilities lacked proper temperature controls. Individual batches tested later were found to have as much as 60 ppm of TCDD.
- 2,4,5-T was withdrawn from use in the USA in 1983, at a time of heightened public sensitivity about chemical hazards in the environment. Public concern about dioxins was high, and production and use of other (non-herbicide) chemicals potentially containing TCDD contamination was also withdrawn. These included pentachlorophenol (a wood preservative) and PCBs (mainly used as stabilizing agents in transformer oil). Some feel that the 2,4,5-T withdrawal was not based on sound science. 2,4,5-T has since largely been replaced by dicamba and triclopyr.
- Agent Orange was a herbicide blend used by the U.S. military in Vietnam between January 1965 and April 1970 as a defoliant. It was a 50/50 mixture of the *n*-butyl esters of 2,4,5-T and 2,4-D. Because of TCDD contamination in the 2,4,5-T component, it has been blamed for serious illnesses in many veterans and Vietnamese people who were exposed to it. However, research on populations exposed to its dioxin

contaminant have been inconsistent and inconclusive. Agent Orange often had much higher levels of TCDD than 2,4,5-T used in the US. The name *Agent Orange* is derived from the orange colour-coded stripe used by the Army on barrels containing the product. It is worth noting that there were other blends of synthetic auxins at the time of the Vietnam War whose containers were recognized by their colours, such as Agent Purple and Agent Pink.

FUNGICIDE

Fungicides are chemical compounds used to prevent the spread of fungi or plants in gardens and crops, which can cause serious damage resulting in loss of yield and thus profit. Though oomycetes are not fungi, they use the same mechanisms to infect plants and therefore in phytopathology chemicals used to control oomycetes are also referred to as fungicides. Fungicides are also used to fight fungal infections.

Fungicides can either be contact or systemic. A contact fungicide kills fungi when sprayed on its surface; a systemic fungicide has to be absorbed by the plant.

The majority of fungicides that can be bought retail are sold in a liquid form. The most common active ingrediant is sulfur, running at 0.08% for the weaker concentrates, and has high as.5% for the more potent fungicides. In powdered form, the concentration is usually around 90%, and is very toxic.

Other active ingrediants in different brands include neem oil, rosemary oil, jojoba oil, and the bacterium *Bacillus subtilis.*

Fungicide residues have been found on food for human consumption, mostly from post-harvest treatments. Some fungicides are dangerous to human health, such as Vinclozolin, which has now been removed from use.

FUNGICIDE RESISTANCE

Pathogens respond to the use of fungicides by evolving resistance. In the field several mechanisms of resistance have been identified. The evolution of fungicide resistance can be gradual or sudden. In qualitative or discrete resistance a mutation (normally to a single gene) produces a race of a fungus with a high degree of resistance. Such resistant varieties also tend to show stability, persisting after the fungicide has been removed from the market. For example sugar beet leaf blotch remains resistant to azoles years after they were no longer used for control of the disease. This is because such mutations often have a high selection pressure when the fungicide is used, but there is low selection pressure to remove them in the absence of the fungicide.

In instances where resistance occurs more gradually a shift in sensitivity in the pathogen to the fungicide can be seen. Such resistance is polygenic – an accumulation of many mutation in different genes each having a small additive effect. This type of resistance is known as quantitative or continuous resistance. In this kind of resistance the pathogen population will revert back to a sensitive state if the fungicide is no longer applied.

Little is known about how variations in fungicide treatment affect the selection pressure to evolve resistance to that fungicide. Evidence shows that the doses that provide the most control of the disease also provide the largest selection pressure to acquire resistance, and that lower doses decreased the selection pressure.

In some cases when a pathogen evolves resistance to one fungicide it automatically obtains resistance to others – a phenomenon known as cross resistance. These additional fungicides are normally of the same chemical family or have the same mode of action, or can be detoxified by the same mechanism. Sometimes negative cross resistance occurs, where resistance to one chemical class of fungicides leads to an increase in sensitivity to a different chemical class of fungicides. This has been seen with carbendazim and diethofencarb.

There are also recorded incidences of pathogens evolving multiple drug resistance – resistance to two chemically different fungicides by separate mutation events. For example *Botrytis cinerea* is resistant to both azoles and dicarboximide fungicides.

There are several routes by which pathogens can evolve

fungicide resistance. The most common mechanism appears to be alternation of the target site, particular as a defence against single site of action fungicides. For example Black Sigatoka, an economically important pathogen of banana, is resistant to the QoI fungicides, due to a single nucleotide change resulting one amino acid (glycine) being replaced by another (alanine) in the target protein of the QoI fungicides, cytochrome b. This presumably disrupts the binding of the fungicide to the protein, rendering the fungicide ineffective.

Upregulation of target genes can also render the fungicide ineffective. This is seen in DMI resistant strains of *Venturia inaequalis*. Resistance to fungicides can also be developed by efficient efflux of the fungicide out of the cell. *Septoria tritici* has developed multiple drug resistance using this mechanism. The pathogen had 5 ABC type transporters with overlapping substrate specificities that together work to effectively pump toxic chemicals out of the cell. Fungi may also develop metabolic pathways that circumvent the target protein, or acquire enzymes that enable metabolism of the fungicide to a harmless substance.

FUNGICIDE RESISTANCE MANAGEMENT

The fungicide resistance action council (FRAC) has several recommended practises to try to avoid the development of fungicide resistance, especially in at-risk fungicides inclding *Strobulins* such as azoxystrobin.

Products should not be used in isolation but rather as mixture, or alternate sprays, with another fungicide with a different mechanism of action. The likelihood of the pathogen developing resistance is greatly decreased by the fact that any resistant isolates to one fungicide will hopefully be killed by the other – in other words two mutations would be required rather than just one.

The effectiveness of this technique can be demonstrated by Metalaxyl. When used as the sole product in Ireland to control potato blight (*Phytophthora infestans*) resistance developed within one growing season. However in countries like the UK where it was only ever marketed as a mixture resistance problems were not seen. Fungicides should only be applied when absolutely necessary, especially if they are in an at-risk group. Lowering the amount of fungicide in the environment lowers the selection pressure for resistance to develop.

Manufacturers' doses should always be followed. These doses are normally designed to give the right balance between controlling the disease and limiting the risk of resistance development. Higher doses increase the selection pressure for single site mutations that confer resistance, as all strains but those that carry the mutation will be eliminated, and thus the resistant strain will propagate. Lower doses greatly increase the risk of polygenic resistance, as strains that are slightly less sensitive to the fungicide may survive.

It is also recommended that where possible fungicides are only used in a protective manner, rather than to try to cure already infected crops. Far fewer fungicides have curative/ eradicative ability than protectant. Thus fungicide preparations advertised as having curative action may only have one active chemical; a single fungicide acting in isolation increases the risk of fungicide resistance. It is better to use an integrative pest management approach to disease control, rather than relying on fungicides alone. This involves the use of resistant varieties and hygienic practises, such as the removal of potato discard piles and stubble on which the pathogen can overwinter, greatly reduce the titre of the pathogen and thus the risk of fungicide resistance development.

Chapter 8

Chemical Fertilizers

Plant nutrition and the soil-plant system. The key-role of fertilizers and their judicious use in crop husbandry is well understood, when one is familiar with the general facts about plant nutrition. It is now known that at least 16 plant-food elements are necessary for the growth of green plants. These plant-nutrients are called essential elements. In the absence of any one of these essential elements, a plant fails to complete its life cycle, though the disorder caused can, however, be corrected by the addition of that element.

These 16 elements are Carbon(C), hydrogen(H), oxygen(O), nitrogen(N), phosphorous(P), sulphur(S), potassium(K), calsium(Ca), magnesium(Mg), iron(Fe), manganese(Mn), zinc(Zn), copper(Cu), molybdenum(Mb), boron(B) and chlorine(Cl). Green plants obtain carbon from carbon-di-oxide from the air; oxygen and hydrogen from water, whereas the remaining elements are taken from the soil.

Based on their relative amounts, normally found in plants, the plant nutrients are termed as macronutrients, if large amounts are involved, and micronutrients, if only traces are involved. The micronutrients essential for plant growth are iron, manganese, copper, zinc, boron, molybdenum, and chlorine.

Most of the plant nutrients, besides carbon, hydrogen and oxygen, originate from the soil. The soil system is viewed by the soil scientists as a triple-phased system of solid, liquid and a gaseous phases. These phases are physically seperable. The plant nutrients are based in the solid phase and their usual pathway to the plant system is through the surrounding liquid phase, the soil solution and then to the plant root and plant cells. This pathway may be written in the form of an equation as

M(Solid)→M(Solution)→N(Plant root)→(Plant top)

where'M'is the plant nutrient element in continual movement through the soil-plant system.

The operation of the above system is dependent on the solar energy through photosynthesis and metabolic activities. This is however, an oversimplified statement for gaining a physical concept of the natural phenomenon, but one should bear in mind that there are many physico and physico-chemical processes influencing the reactions in the pathway. The actual transfer in nature takes place through the charged ions, the usual form in which plant-food elements occur in solutions(liquid phase of the system).

Plant roots take up plant-food elements elements from the soil in these ionic forms. The positively charged ions are called'cations'which include potassium(K^+), Calcium(Ca^{++}), magnesium(Mg^{++}), iron(Fe^{+++}), zinc(Zn^{++}), and so on. The negatively charged ions are called anions and the important plant nutrients taken in this form include nitrogen(NO-3), phosphorous($H_2PO^-_4$), sulphur(SO^-_4), Chlorine(Cl), etc.

The process of nutrient uptake by plants refers to the transfer of the nutrient ions across the soil root interfaces into the plant cell. The energy for the process is provided by the metabolic activity of the plant and in its absence no absorption of nutrients take place. Nutrient absorption involves the phenomenon of ion exchange. The root surface, like soil, carries a negative charge and exhibits cation-exchange property. The most efficient absorption of the plant nutrients takes place on the younger tissues of the roots, capable of growth and elongation.

In this respect, root-systems are known to vary from crop to crop. Hence their feeding power differs. The extent and the spread of the effective root-system determines the soil volume trapped in the feeding-zone of the crop plant. This is indeed an important information in a given soil-plant system which helps us to choose fertilizers and fertilizer-use practices. The absorption mechanisms of the crop plants are fairly known now. There are three mechanisms in operation in the soil-water-plant systems. They are

- The contact exchange and root interception,
- The mass flow or convection,
- Diffusion.

In the case of contact exchange and root interception, the exchangeable nutrients ions from the clay-humus colloids migrates directly to the root surface through contact exchange when plant roots come into contact with the soil solids. Nutrient absorption through this mechanism is, however, insignificant as most of the plant nutrients occur in the soil solutions. Scientists have found that plant roots actually grow to come into contact with only 3 percent of the soil volume exploited by the root mass, and the nutrient uptake through root interception is even still less.

The second mechanism is mass flow or convection, which is considered to be the important mode of nutrient uptake. This mechanism relates to nutrient mobility with the movement of soil water towards the root surface where absorption through the roots takes place along with water. Some are called mobile nutrients. Others which move only a few millimetres are called immobile nutrients. Nutrient ions such as nitrate, chloride and sulphate, are not absorbed by the soil colloids and are mainly in solution.

Such nutrient ions are absorbed by the roots along with soil water. The nutrient uptake through this mechanism is directly related to the amount of water used by the plants (transpiration). It may, however, be mentioned that the exchangeable nutrient cations and anions other than nitrate, chloride and sulphate, which are absorbed on soil colloids are in equilibrium with the soil solution do not move freely with water when it is absorbed by the plant roots.

These considerations, therefore, bring out that there are large differences in the transport and root absorption of various ion through the mechanism of mass flow. Mass flow is, however, responsible for supplying the root with much of the plant needs for nitrogen, calcium and magnesium, when present in high concentrations in the soil solution, but does not do so in the case of phosphorous or potassium. The nutrient uptake through mass flow is largely dependent on the moisture status of the soil and is highly influenced by the soil physical properties controlling the movement of soil water.

The third mechanism is diffusion. It is an important phenomenon by which ions in the soil medium move from a point of higher concentration to a point of lower concentration. in other words, the mechanism enables the movement of the nutrients ion without the movement of water. The amount of nutrient-ion movement in this case is dependent on the ion-concentration gradient and transport pathways which, in turn, are highly influenced by the content of soil water. This mechanism is predomionant in supplying most of the phosphorous and potassium to plant roots. It is important to note that the rhizophere volume of soil in the immidiate neighbourhood of the effective plant root receives plant nutrients continously to be delivered to the roots by diffusion. However, when the nutrient concentration builds up far excess of the plant in the reverse direction. These are some of the choice of fertilizers and fertilizer practices for practising scientific agriculture.

The relationship in the soil-plant system stated in the simple equation give in the earlier paragraph reflects the highly dynamic nature of the soil solution. One knows that the roots of the growing plants continuously remove nutrient ions from the soil solutions. At the same time, the breakdown of the soil minerals and the generating of more exchangeable cations, the biological activity and the additions made to the anions, e.g. nitrates, continuously change the composition of the soil solution. At a given point of time, therefore, the available plant nutrients in the soil solution may range from a tiny amount to larger quantities. Under favourable conditions, crop plants, in general, require larger amounts of plant nutrients than the quality found in soil solution at any given time. Hence, the situation of nutrients supply to plants becomes a limiting factor, specially, at the critical stages of plant growth and low crop yeilds result in recognition, therefore, fertilizers application and the use of suitable fertilizers are recommended for higher crop yeilds in productive farming. The knowledge of the specific role of each essential element in the growth of crop plants and their amounts required for efficient crop production is considered necessary in adopting scientific fertilizers use.

PLANT NUTRIENTS AND THEIR FUNCTIONS

The plants require, the following essential nutrients for their normal development

Carbon	Nitrogen	Calcium
Hydrogen	Phosphorous	Magnesium
Oxygen	Potassium	Sulphur
Iron	Zinc	Chlorine
Manganese	Boron	—
Copper	Molybdenum	—

Carbon is obtained from carbon dioxide of the air; Oxygen from air and water; Hydrogen from water; Nitrogen from air and soil or both, and all other nutrients from the soil. Soil is a the most important source of plant food. Nitrogen, Phosphorous and Potassium are known as primary plant nutrients; Calcium, Magnesium and Sulphur are secondary nutrients; Iron, Manganes, Copper, Zinc, Boron, Molybdenum and Chlorine as trace elements or micronutrients. The primary nutrients and secondary nutrients elements are known as major elements. This classification is based on their relative abundance, and not their relative importance. the micronutrients are required in small quantities, but they are as important as the major elements in plant nutrients.

Air is the primary source of Nitrogen for plant nutrient. Only leguminous crops can directly use this free Nitrogen with the help of symbiotic bacteria of the genus Rhizobium. Other plant derive from soil their Nitrogen in the form of Nitrogen and Ammonium. Nitrogen and Ammonium are produced in the soil by action of micro-organisms on the soil organic matter. Non symbiotic micro-organisms can fix free Nitrogen of the air and make it available to plant in Ammonium and Nitrates forms.

Nitrogen encourages the vegetative development of plants by importing a healthy green colour to the leaves. It also controls, to some extent the efficient utilization of phosphorous and Potassium. Its dependency retards growth and root development, turns the foilage yellowish or pale green, histens maturity, causes the shrivelling of grains and lowers crop yield. The older leaves are affected first. An excess of Nitrogen produces leathery(sometimes crinkled), dark-green leaves and succulent growth. It also delays the maturation of plants, impairs the quality of crops like barley, potato, tobacco, sugarcane, and fruits; increases susceptibility to diseases and causes'lodging'of cereal crops by inducing an undue lengthoning of the stem internodes.

Phosphorous influences the vigour of plants and improves the quality of crops. It encourages the formation of new cells, promotes root growth(particularly the development of fibrous roots), and hastens leaf development through emergence of ears, the formation of grains, and the maturation of crops. It also increases resistence to diseases and strengthens the stems of cereal plants, thus reducing their tendency to lodge. It offsets the harmful effects of excess nitrogen in the plant. When applied to leguminous crops, it hastens and encourages the development of nitrogen-fixing nodule bacteria. If phosphorous is deficient in the soil, plants fail to make a quick start, do not develop a satisfactory root-system, remain stunted and sometimes develop a tendency to show a reddish or purplish discolouration of the stem and foilage owing to an abnormal increase in the sugar content and the formation of anthoscyanin.

However, the deficiency of this element is not easily recognised as that of nitrogen. It has also been observed that cattle feeding on the produce of deficient soils become dwarfed, develop stiff joints and lose the velvetty feel of the skin. Such animals show an abnormal craving for eating bones and even soil itself.

Potassium enhances the ability of plants to resist diseases, insect attacks, and cold and other adverse conditions. It plays an essential part in the formation of starch and in the production and translocation of sugars, and is thus of special value to carbohydrate-rich crops, e.g. sugarcane, potato and sugar-beet. The increased production of starch and sugar in legumes fertilized with potash benefits the symbiotic bacteria and thus enhances the fixation of nitrogen. It also improves the quality of tobacco, citrus etc. With an adequate supply of potash, cereals produce plump grains and strong straws. But an excess of element tends to delay maturity, though, not to be the same extent as nitrogen.

Plants can make up and store potassium in much larger for correcting zinc deficiency. The symptoms of zinc deficiency appear generally in younger leaves, starting with intervienal chlorosis leading to a reduction in shoot growth and the shortening of internodes. Mottle leaf, little leaf, etc. in the case of trees are symptoms of zinc difficiency. The buds of several defficient maize plants become white; in citrous interveinal chlorosis and mottled leaf occur. In calcareous soils and in soils with very high phosphorus content, zinc defficiency is commonly expected to occur. The principal function of zinc in plants is as a metal activator of enzymes.

In highly weathered coarse textured soils zinc deficiency appears under an intensive cropping programme. The availability of zinc is least between pH 5.5 and 7, but its availability increases at a lower pH. At a higher pH above 7, zinc availability becomes a complex problem, as the positively charged zinc ion gets converted into a negatively charged zincate complex whose availability tends to be reduced in alkaline soils.

When the calcium ion is predominent, the highly insoluble calcium zincate is formed and zinc availability gets seriously limited. The application of soluble zinc salts or zinc chelates to the soil is generally recommended to correct its defficiency. Foliar sprays are advocated, especially for orchard trees for amending zinc defficiency. About 5 to 50 kg of zinc sulphate per hectare is used for such purposes.

The symptoms of boron defficiency vary with the kind and age of the plant, the conditions of growth and the severity of the deffiency. Each crop produces its characteristic growth abnormalities associated with boron defficiency, such as yellows and rosetting in lucerne, snakehead in wallnuts, die-back and corking of fruits in apple, corking and pitting of fruits in tomatoes, hollow stem and the bronzing of curd in cauliflower, the brown-heart diseases in table-beets, turnips, etc.

Molybdenum deficiency produces whip-tail in cauliflower, broccoli and other Brassica crops. The deficiency of this element reduces the activity of the symbiotic and non-symbiotic nitrogen-fixing micro-organisms. It was in 1954 that chlorine was proved to be an essential micronutrient. Its defficiency under field conditions has not been reported so far. In water-culture solutions, the leaves of chlorosis, necrosis and an unusual bronze discolouration on tomatoes.

Sodium is not an essential element for plant growth. But some crops, such as beet, celery, cabbage, kale, knol-khol, radish, rape and turnip, benefit greatly by application of soluble sodium salts, specially if the soil is deficient in potassium. Sodium is also of direct benefit to plants indigneous to the sea-shore or to irrigated arid regions. Salts of this element are said to release more of potassium from the exchange complex and to help to maintain phosphorus in a more available form. They also serve as a partial substitute for potassium in the case of potatoes and cotton.

MAINTENANCE OF SOIL FERTILITY

No two soils are alike either in respect of their nature or in respect of quantities of plant nutrients they contain. Under a given situation, the system of farming, soil management and manuring practices, etc., influence the productiovity of soils and crop yields obtained from them. The quantities of the three primary nutrients- N, P_2O_5 and K_2O removed from a hectare of land by some of the important crop.

It is estimated that the different agricultural crops in India remove about 4.27 million tonnes of nitrogen, 2.13 million tonnes of phosphoric acid, 7.42 million tonnes of potash and 4.88 million tonnes of lime per year. The production of larger yields through improved varities of crops and intensive cultivation will increase the depletion of nutrients still further. But erosion and leaching cause additional losses. The present production of synthetic nitrogenous fertilizers in the country reached 1.5 million tonnes of nitrogen in 1975-76, and the bulky organic manures might supply another 1.5 million tonnes of total nitrogen.

The amounts of phosphorous, potash, etc., added to the soil are very small. It is thus obvious that the current huge drain on nutrient supplies will continue to impoverish the soils unless these supplies are replenished by natural or by artificial means, the principal methods of supplementing natural recuperation and for improving the productive capacity of the soils are

- To add organic matter to the soil, so that through decay, it may furnish a more or less continuous supply of nuttrients for crops,
- To restore or increase the amount of defficient nutrients by the application of fertilizers.

The urgent need of the constantly expanding agriculture production to meet the requirements of the continually increasing human and cattle populations in India makes the supply of additional plant nutrients through fertilizers and organic manures, a problem of supreme importance.

Table The Quantities of Plant Nutrients Removed from soil by Different Crops (Kg/ha)

Crop	Yield (grain)kg/ha	N	P_2O^5	K_2O
Rice	2,240	34	22	67
Wheat	1,568	56	24	67
Jawar	1,792	56	15	146
Bajra	1,120	36	22	66
Maize	2,016	36	20	39
Barley	1,120	41	20	35
Sugarcane	67,200	90	17	202
Groundnut	1,904	78	22	45
Mustard	672	22	11	28
Linseed	1,008	19	12	33
Cotton	448	30	17	45
Jute	1,568	67	34	67

Tea	896	45	13	28
Coffee	896	34	11	34
Tobacco	1,456	94	57	91

TYPES OF MANURES AND FERTILIZERS

Indian soils are usually very poor in organin matter as well as in nitrogen. Pohsphate deficiency is less wide-spread and potash deficiency generally occurs in com-pact areas. In acid soils the addition of lome steps up production.

Materials which are commonly used to maintain and improve soil fertility may be classified as follows

- *Manures:* These are relatively bulky materials, such as animal or green manures, which are added mainly to improve the physical condition of the soil, to replenish and keep up its humus status, to maintain the optimum conditions for the activities of soil micro-organisms and make good a small part of the plant nutrients removed by crops or otherwise lost through leaching and soil erosion. They, thus, supply practically all the elements of fertility which crops require, though not in adequate proportions. The plant-food elements contained in a manure are released in an available form after it is applied to the soil and is decomposed by soil micro-organisms. Similarly, the green manures add not only substantial amounts of organic matter but also nitrogen.
- *Fertilizers:* Fertilizers are inorganic materials of a concentrated nature; they are applied mainly to increase the supply of one or more of the essential nutrients, e.g. nitrogen, phosphorous and potash. Fertilizers contain these elements in the form of soluble of readily available chemical compounds. This distinction is, however, not very rigid. In common parlance the fertilizers are sometimes called'chemical','artificial'or'inorganic'manures.
- *Concentrated organic manures:* Some of the concentrated materials, such as oil-cakes, bone-meal, urine and blood are of organic origin. The use of manures and

fertilizers is complementary and not as a substitute for each other.

- *Bulky Organic Manures:* The properties and role of organic matter and humus in the soil have been explained already. The average nutrient contents of manures and other organic raw materials which may be used to maintain the humus content of the soil.

FARMYARD MANURE

Good-quality farmyard manure is perhaps the most valuable organic matter applied to a soil. It is the most commonly used organic manure in India. It consists of a mixture of cattle dung, the bedding, used in the stable and of any ramnants of straw and plant stalks fed to cattle. Though its crop-increasing value has been recognised from time immemorial, more than 50 per cent of the cattle dung produced in the country today is burnt as fuel and is thus lost to agriculture. Not only this tremendous waste, but also the tradition method of preparing and storing the farmyard manure is generally faulty.

The cattle-dung, together with stable-waste and house sweeping, is first collected in the open backyard, and when a cartload has been collected, it is removed to another heap or to an uncovered pit in a common plot outside the village. The loose heaps lie exposed to the sun, with the result that the raw organic matter dries up quickly and does not rot properly. Very often, a part of the dry dung is blown off by wind or washed away by rain.

Cattle urine is either not conserved or is stored in a defective manner. American studies on the distribution of soil derived elements between urine and faeces of dry cows have shown that 95 per cent pf Potassium, 63 per cent of nitrogen and 50 per cent of sulphur are contained in the urine. The wastage of nitrogen-rich urine, the loss of nitrogen(in the form of ammonia) due to the fermentation of exposed cattle dung, and the washing away of soluble mineral elements be leaching reduce its manurial value in India to a great extent.

About half of this nitrogen, one-sixth of phosphorous and more than half of potash are readily soluble and subject to dissipation. However, the loss of nitrogen and minerals elements caused by careless handeling can be reduced greatly by using absorbent bedding for cattle, storing dung in stone or brick-line pits, mixing large quantioties of straw and other vegitable matter with cattle dung, and keeping the heap compact and moist. Thus, if urine is properly conserved, the loss of soluble mineral elements through seepage is prevented, bacterial decomposition of raw organic matter is encouraged, plant nutrients are made soluble, and nitrogen losses are minimised.

If urine is not conserved in the bedding used for cattle it must be collected in covered pucca cistern and then, added to the dung in the manure pit. Nitrogen in the urine is mainly in the form of urea, which readily changes into the highly volatile ammonium carbonate through bacterial action, and quickly loses ammonia thereafter by evapouration. This loss can be reduced to a great deal if the manure and the urine-soaked absorptive itter for bedding are kept compacted in a pit. The pit may be 1 m in depth, 1.3 to 1.5 m in width and 4.5 to 6 m in length, depending upon the no. of cattle on a farm.

The filling of the pit should be'sectional'and when each section of three or 1.3 m in length is filled to about 45 cm above the ground level, it should be clustered with 2.5 cm layer of a mixture of mud and dung in equal proportions. Before plastering, 4 to 5 buckets of water should be added to the manure in the pit. Plastering conserves moisture and nitrogen and also prevents housefly nuisance. The manure becomes ready for use in about 4 to 5 months after plastering.

The quality of manure is also improved by the concentrated feeds given to cattle. Cotton-seed, cotton-seed cake, linseed-meal, wheat bran, grain husk, groundnut cake, gram, horse-gram, etc. are rich in nitrogen, phosphorous, potassium, magnesium and sulphur. It has been found that in the case of adult working-cattle about 80 per cent of nitrogen and the other mineral elements contained in the feed is recovered in urine, faeces and other animal by-products.

Accordingly, manure from cattle fed on cereal straws and grass hay is much less valuable than that from animals fed on legume hays, grains and concentrates.

In foreign countries, considerable attention has been given to the use of preservatives on manure. Calcium sulphate or gypsum and superphosphate have proved most promising in preventing the escape of ammomnia. Gypsum has been found specially effective as an ammonia-absorbing agent. Superphosphate, besides absorbing ammonia, supplies additional phosphorous and, thus, improves the crop producing capacity of the manure.

Partially rotten farmyard manure should generally be applied to the soil about three to four weeks before the sowing of a crop. In case there is sufficient moisture in the soil, there sill be enough time for its decomposition and for improving the soil structure. Its application too long before sowing a crop will either cause a drying up of the rotted manure or too quick decomppostion, depending on the incidence of rains. But in each case, there will be a serious loss of ammonia and nitrogen. If the manure is already well rotted, it is advisable to apply it just before sowing to a crop.

This procedure is perticularly essential in the case of light soils. In any case, after the manure is carted to the field, it should be evenly spread and worked into the soil soon to avoid the loss of nitrogen. The existing practice of leaving the manure in small heaps scattered in the field for several days, before spreading it in the field and incorporating it into the soil, results in the serious deterioration of its quality, particularly if strong winds blow. In vagitable and fruit cultivation, the application of well-rotten manure, in conjunction with fertilizers to young plants individually, has been founnd to give the best results. In Egypt, even the cotton crop, which is invariably grown on ridges, is given a top-dressing a handful of the rotten manure being applied to the soil at the base of each plant before an irrigation, and is worked into the soil with a hand hoe.

The practice of penning cattle, sheep and goats in the fields in summer is common in some parts of the country. Folding

7,000 sheep for one night is said to add the equivalent of 149.3 quintals(14.93 tonnes) of cattle dung. The fresh dung left in the field in such cases rapidly dries up. This drying checks ammonification and loss of nitrogen. With the first fall of rain, the dung is worked into the soil. It, therefore, does not lose much of its fertilizing value. Further more, its beneficia effects on the physical condition of the soil are undeniable. However, sheep-folding is said to make the land more weedy.

There must be adequate moisture in the soil for the proper decomposition of organic matter. Farmyard manure can, therefore, be applied to all crops grown in the rainy season or grown under irrigation. The quantity of manure to be applied to unirrigated crops varies from 1.5 to 2 cartloads per hectare in areas of heavier rainfall. If sufficient farmy manure is not available, it may be applied at the usual rate to a part of the land,say, to one-third or one-forth of the area, in rotation every year, so that all parts of the field receive the manure regularly once in three or four years. For irrigated field crops, the rate varies from 10 to 20 cartloads. Sugarcane, maize and garden crops, such as potatoes, turmeric, ginger, vegetables and fruits receive still higher doses, amounting sometimes to 15 to 25 cartloads. A cartload of manure, measuring 9 cubic metres, weighs about half a tonne. It must be stressed that the value of farmyard manure in soil improvement is due to its content of principal nutritive elements and its ability to

- Improve the soil tilth and aeration,
- Increase the water-holding capacity of the soil,
- Stimulate the activity of micro-organisms that make the plant-food elements in the soil readily available to crops.

The supply of organic matter, which is later converted into humus is a property of farmyard manure.

One tonne of carttle dung can supply only 2.95 kg of nitrogen, 1,59 kg of phsophoric acid and 2,95 kg of potash. The use of farmyard manure alone causes an imbalance in nutirtion owing to its relatively low content of phosphoric acid. Therefore, to keep the soils well supplied with all the essential elements of plant food in a readily available form, and also to keep them in good'heart', it is advisable to use the bulky organic manures in conjuction with superphosphate and such other artificial fertilizers as contain the particular plant food or foods in which a soil may be deficient or which the crop to be grown may specially require.

COMPOSTED MANURE

Another method of augmenting the supplies of organic is the preparation of compost from farmhouse, and cattle-shed wastes of all types. Composting has been advocated and adopted extensively during the past 25 years. Composting is the process of reducing vegetable and animal refuse(rural or urban) to a quickly utilizable condition for improving the maintaining soil fertility. Research conducted in India and abroad has shown that good organic manure similar in appearance and fertilizing value to cattle manure similar in appearance and fertilizing value to cattle manure can be produced from waste materials of various kinds, such as cereal straws, crop stubble, cotton stalks, groundnut husk, farm weed and grasses, leaves, leaf-mould, house refuse, wood ashes, litter, urine-soaked earth from cattle-sheds and other similar substances. These raw vegetables materials are rich in cellulose and other readily decomposable carbohydrates and have a carbon-nitrogen ratio of 40 or more to 1.

The direct application of such undecomposed, low nitrogen organic matter as manure brings about a temporary deficiency of mineral nutrients(specially nitrates and ammonium compounds) in the soil by stimulating the growth of micro-organisms, which in turn, compete with crop plants for available nitrogen, phosphorous and other elements. Hence, before using them as manure, it is necessary to compost or partially decompose them. This process lowers the carbon-nitrogen ratio to about 10 or 12 to 1.

Two methods of composting waste organic materials are usually recommended. One depends on aerobic and other one unaerobic decomposition. In both cases, the farm wastes have to be used as a bedding for cattle in order to absorb a large of the animal urine. In the aerobic process, the used bedding, the sweepings from cattle sheds and some urine-soaked earth from the stable floar are removed everyday, mixed with a little cattle dung and 2 or 3 handfuls of wood ashes are deposited on a well drained site to gradually builed up a lowpile about 30 or 45 cm in height, about 5 m in width and of any convenient length.

The pile is built up before to start of rainy season. After the first heavy shower, the waited material in 1.2 strip on each side of the long heap is turn with a rake on to 2.4 m wide stripe in the middle, so raising the height of the heap to nearly 1 m. This process prevents loss of misture and ensures a quick start of decomposition. When the heap sinks appreciably and such a sinking takes about 3 to 4 weeks, it is given a turning and made into fresh heap, thus mixing the outside material with that from inside. After about a month or more, depending on incidence of rains the heap is given a final turning on a cloudy or moderately rainy day and rebuilt the vacant part of the original position. The composed becomes ready for use in about 4 months. This method is eminently suited for composting in the rainy season.

In Tamil Nadu, the application of 90 kg of partially fermented dung in the form of a thin suspension, with 22.50 kg of bone-meal per tonne of dry material to be composted is recommended. If urine-soaked earth and cattle-dung are not available, the raw organic materials can still be composted, provided more than one-third of the residues is soft and finely broken, such as fallen leaves, leaf-mould, kitchen wastes, grass clippings, green and succulent weeds, plant trimming and clopped wheat, barley and soft cereal straw. The use of ordinary soil and wood ashes or lime is, however, essential.

In the anaerobic process, the mixed farm residues are collected in pits of a convenient size, say, 4.5 m*1.5 m*1 m. Each day's collection is spread in a thin layer, sprinkled with a mixture of fresh cowdung(4.5 kg), ashes(140 to 170 g) and water (18 to 22 litres) and compacted. The pit is filled till the raw material stands 38 to 46 cm above its edge, and then is then plastered with a 2-5 cm layer of a mixture of mud cowdung. Under such conditions, decompostion is anaerobic and high temperatures do not develop. Insoluble nitrogen compunds gradually become soluble and the carbonaceous matter is broken down into carbon dioxide and water.

The loss of ammonia is negligible, because in high concentrations of carbon dioxide, ammonium carbonate is stable. The plastered pit also prevents the fly nuisance. The well-made compost contains 0.8 to 1 per cent of nitrogen and has all the good properties of farmyard manure. It can be used in the same way as the latter. The anaerobic process is particularly suited for use by gardeners in or near cities and towns.

COMPOSTING METHODS

RAW MATERIALS

The materials are needed are mixed plant residues, animal dung and urine,earth, wood ash and water. All vegetables wastes available on a farm like weeds,stalks,stems,fallen purnings,chaff,fodder remnants,green matter and so on, are collected and stacked in a pile. Hard woody material like cotton or pigeon-pad stalks and stubble are first spread on the farm road and crushed under Vehicles viz. tractors or bullock carts before being piled. Such hard materials should in any case not exceed 10 percent of the total plants residues.

Green material which are soft and sufficient and allowed to wilt for 2 or 3 days to remove access moistures before stacking; they tend to pack closely, if they are stacked in the fresh state. While stacking, each material is spread in layers of about 15cm thickness until the heap is about 1m and 50cm high.The mixture of different kinds of vegetable residuesensures a more efficient decomposition. The heap is then cut into vertical slices and about 20-25 kg is put under the feet of cattle in the shed as bedding for the night. The next morning the bedding, along with the dung and the urine, and urine earth, is taken to the pits where the composting is to be done.

PIT METHOD

This method includes following steps

Site and Size of Pit

The site selected for the compost pit should be near to cattle shed and water source at high level so that no rain water gets in during the mansoon season. A temporary shed amy be constructed over it to protect the compost from heavy raifall. The pit should be about 1 m deep * 1.5-2.0 m wide and of any suitable length.

Filling the Pit

The material brought from the cattle shed is spread and on each layer is spread a slurry of dung made with 4.5 kg urine earth and 4.5 kg of inoculum taken from a 15-day-old composting pit. A sufficient quantity of water (nearly 90%) is sprinkled over the material in the pit to the wet it. The pit is filled in this way layer-by-layer and it should not take longer than 1 week to fill. Care should be taken to avoid compacting the material in any way.

Turning

The material is turned 3 times during the whole period of composting (i) after 15 days from filling the pit, (ii) another 15 days, (iii) after another 30 days. At each turning the material is mixed throughly, moistened with water and replaced with the pit.

HEAP METHOD

During rainy seasons or in regions with heavy rainfall the compost may be prepared in heaps above ground. when sufficient nitrogeneous material is not available a green manure or leguminous crop like sunhemp is grown on the fermenting heap by sowing seeds after the first turning. The green mater is then turned in at the second mixing.

Dimensions

The basic Indore pile is about 2 m wide at the base, 1.5 m high 2 m long. The sides are tapered so that the top is about 0.5 m narrower in width than the base. A small bund is sometimes built around the pile to protect it from wind which tends to dry the heap.

Forming the Heap

The heap isusually commenced with a 20 cm layer of carbonceous material such as leaves, hay-straw, sawdust, wood chips and chopped corn stalks. This is then covered with the 10 cm of nitrogeneous material such as fresh grass, weeds or garden plant residues, garbage, fresh or dry manure or digested sewage sludge. The pattern of 20 cm carbonaceous material and 10 cm nitrogeneous material is followed until the pile is 1.5 m high and they are normally wetted so that they feel damp but not soggy. The pile is sometimes covered with soil or hay to retain heat and is turned at 6- and 12-weeks-interval. In the republic of korea, heaps are covered with thin plasic sheets to retain heat and it has also resulted in the death of insects.

If materials are limited, the alternate layres can be added as they become available. Also, all materials may be mixed together in the pile, if one is careful to maintain the proper proportions. Shredding the material speeds up deconposition considerably; most materials can be shredded by running over it several times with a rotary-mower.

Advantages and Limitations

Preparation on a large scale can be done through community composting. there is lack of protection from rain and wind. A considerable amount of water is needed ans so the heap method is not suitale in near areas of scanty rainfall. The intense aerobic decomposition to which the material is subjected no doubt shortens the period of composting but it leads to heavy losses of organic matter and nitrogen. Therefore the C N ratio should be maintained between 30 and 40 to reduce such losses.

BANGALORE METHOD

Preparation of the Pit

Trenches or pits 1m deep are dug; the breadth and length of the trenches can be made depending on the availability of land and the type of material to be composted. The selection of site for 1 pit is made as mentioned in the Indore method.

The trenches should preferably have sloping walls and a floor of 90 cm slope to prevent waterlogging.

Filling the Pit

Organic residues and night soil are put in alternate layers and after filling, the pit is covered with a 15-20 cm thick layer of refuse. the materials are allowed to remain in the pit without turning and watering for 90 days. during this period the material settles down due to reduction in volume of the biomass and additional nightsoil and refuse in alternate layers are placed on top and plastered or covered with mud or earth to prevent loss of moisture and breeding of files. the material undergoes anaerobic decomposition at a vey slow rate and it takes about 180-240 days to abtain the finished product.

ADVANTAGES AND LIMITATIONS

The recovery of the finished product is greater as compared to aerobic composting but loss of nitrogen is negligible. Labour requirements is less than for the Indore method as turning of material is not done; labour is needed only for digging and filling the pits. The methods requires along time to produce a finished compost and so takes up more land use. A uniform high temperature is not assured in the biomass. Problems of odour and fly breeding need to be attended too.

SYNTHETIC COMPOST

In the preparation of synthetic compost, the organic nitrogen as dung, required by micro-organism, can be completely substituted by inorganic nitrogen compounds like ammonium sulphate and urea which are utilized equally effectively for decomposition of carbonaceous materials into compost. The Adco process of preparations of synthetic compost developed by Hutuchinson and richards is based on this principle. This facilitates the utilization of large quantities of various organic waste material where supplies of dung are either short of the requirement or not available at all, as on mechanized farms.

The basic principle of C N ratio in manure preparation can be applied to add nitrogenous fertilizers in sufficient quantity to decompose. The material to be composted is moistened. This is sprinkled with the fertilizer solution and then with lime. Superphosphate may be added to fortify the phosphorous content of the manure. The treatment is continued layer-wise until the heap or pit is filled to the size and allowed to ferment. The manure becomes ready for application in about 120-180 days and resembles with farmyard manure in its action on soil and plant growth.

LEAF COMPOST

Leaf composting, can be achieved by heap or ditch composting or by windrow composting. windrow are preferred as they allow efficient handling of materials Provide good aeration, allow sufficient of water and are easy to be performed.

It is suggested the formation of uniform-shaped windrow from 2.40 m-3.60 m at the base and 2.40m-3.00m high and of any convenient lenght. Windrow built too high will have excessive compaction at the base resulting in anaerobic conditions. Windrows built too low will not allow sufficient insulation to sustain thermophilic temperatures during cold weather. To ensure proper aeration, it is important to break apart tightly compacted leaves.

Though reduction in size may aid in rapid decomposition, it is not desirable in leaves because it increases the compaction making more frequent aeration necessary. When the incoming leaves are not adequately moist, it is desirable to add water to maintain a proper moisture regime of 40 to 60%.

The C N ratio of leaves is relatively high. It can be as high as 80, and needs to be amended with nitrogen. Sewage sludge, urea and grass clippings are good sources of nitrogen. If a nitrogen source is to be added, caution should be exercised to distribute it uniformly throughtout the windrow, lest it may result in undesirable anaerobic conditions and uneven decomposition. Proper aeration can be maintained by periodical mixing of the material.

Under optimum environmental composting conditions, leaf compost will be ready between 180 and 270 days. Leaf mould will have final PH range of 6-7. It is suggested the use of finished compost (leaf mould) as converting material (10-15cm) in the subsequent preparation of leaf compost to supply a heavy inoculum of micro-organisms.

ENRICHMENT WITH PHOSPHOROUS

Phosphorous enriched compost is prepared by adding 5% superphosphate at the filling of the compost pits. Other sources of phosphorous for this purpose are powered rock phosphate, preferably of low grade (less than 11% P), can be used with profit. Besides, phosphorous it is a source of calcium and micronutrients. Bonemeal will provide nitrogen as well as phosphorous; it contain 9-11% P and 2-4% N. Steamed bonemeal is more easily ground than the fresh material. It contains a little less nitrogen but more phosphorous than the raw material. Basic slag provides calcium, magnesium and trace nutrients and small amount of phosphorous. Banana residues contains about 1-1.5% phosphorous on an ash basis.

ENRICHMENT WITH POTASSIUM

Granite dust of powdered potassium-containing minerals like feldspars can be added to enrich compost. Potassium and other deficient elements can be added to compost by including plant materials which contain apperciable amounts of those elements. Water hyacinth for example, is a rich source of potassium and of many other elements required by plants. Banana skin and stalks contain 34-42% potassiumon an ash basis, seaweeds are rich in iodine, boron, copper, magnesium, calcium and phosphorous. Leaves are also a good source of trace elements and should form a parta part of every compost heap. Potato peel is rich in trace elements and dry potato vines contain 1% potassium, 4% calcium and 1% magnesium.

VERMI COMPOSTING

VERMI-COMPOSTING is the use of earthworms for composting of organic residues. Earthworms can consume practically all kinds of organic matter. One worm, weighs about 0.5 to 0.6 g, eates wastes as their own body weight perday and produces cast of the same weight per day. It is estimated that 1000 tonnes of moist organic matter can be converted by earthworms into 300 tonnes of compost. Organnic materials undergo comples biochemical changes in the intestines and vermi-composting is an appropriate technique for disposal of non-toxicsolid and liquid organic wastes. It helps in cost effective and efficient recyclicing of animal wastes (poultry, equine, piggery excreta and cattle dung) agriculture residue and industrial wastes using low energy, excreta together with their cocoons and undigested feed make up vermicasting. The castings of earthworms are rich in nutrients (N, P, K, Ca and Mg), and also in bacterial and actinomycetes population.

The actinomycetes population in worm cast is over 6 times more than in the original soil. A mosist compost heap (30-40% moisture level) of 2.4 m * 1.2 m * 0.6 m high, can support a population of mare than 50,000 worms. The temperature is the culture bed should be within the range of 200-300C. The introduction of worms into compost heap has been found to mix the materials, aerate the heaps and hasten decomposition. turning the heaps is not necessary, if earthworms are present to do the mixing and aeration. Besides rural and urban wastes, effluents from agro-industries viz dairies, tanneries, pulp and paper mills, distilleres etc.can be treated by using earthworms.

Benefits of Earthworms

Earthworms help in the preparation of compost maintaining soil health as follows

- Improvement in fertility of soil.
- Arnelioration of physical condition of soil.
- Mixing of sub-soil and top soil.
- Correction of undetermined deficiences in plants.
- Use of earthworms in recycling of city and rural wastes, sewage waste waters and suldge, and industrial wastes eg. paper, food and wood industries.
- Supplementing traditional feeds

SPECIES FOR VERMI-COMPOSTING

Earthworm can be divided as surface living (epigeic) and burrowing (epianecic) worms. Epigeic or compost worms are found on surface and are reddish brown eg. Lumbricus rubellus (red worms). Of many species of earthworms tested for mass culture all over the world, Eisenia fetida, Eudrilus eugeniae and Perionyx excavatus come in the above order of preference for their ability to compost organic wastes. The shapes of cocoons of Eisenia frtida and Eudrilus eugeniae are dissimilar.

REARING OF EARTHWORMS

The worms are reared and multiplied from a commercially obatined breeder stock in shallow wooden boxes of 45 cm *60 cm, provided with drainage holes and stored on shelves and tiers. A bedding material is compounded from miscellaneous organic residues saw dust, cereal straw, rice husks, sugarcane trash, bagasse, paper, cardboard, coir waste, grasses etc. and is moistened well with water. The wet mixture is stored for 30 days covered with a damp sack and is throughly mixed several times. When fermentation is complete, chicken manure and green matter eg. Leucaena leaves or water hyacinth is added. The material is placed in the boxes in the ans sufficiently loose for the worms to burrow and should be able to retain moisture. the The proportion of the different materials will vary according to the nature of the material but a final nitrogen content of about 2.4% should be amied at.

A pH value as near neutral as possible is necessary and the boxes should be kept at temperatures between 200C and 270C. At higher temperatures the worms will aestivate and at lower temperatures they hibernate. for each 0.1 m2 of surface area 100 g of breeder worms are added to the boxes. Inspite of their being able to eat the bedding material, the worm at this stage are regularly fed @ 1 kg of feed a day for every kg of worms. the feed stuffs used are again various types of organic matter and include partially digested cowdung, chicken manure, Leucaena leaves, vegetable waste and water hyacinth.

Some form of protection against predators like birds, vats, ants, frogs, leeches and centipeded is provided to the worms.

VERMI-COMPOSTING IN PITS

A number of pits 2m *1m with sloping sides aredug having suitable dimension. Vermi-composting is done in pits and in vitro. Both of these are discussed here.

Bamboo poles are laid in paralled row on the pit's. Its floor is with a lattice of wood strips. Necessary drainage is provided because worms can not survive in a waterlogged condition. Alternatively to this and sand can be placed in the bottom of the pit to facilitate proper drainage. Above this a thick layer (15-20 cm) of good loamy soil should be spread. The pit can now be filled with available organic residues such as animal manure, leaves and green weeds, crop residues etc. Moisture levels of the contents of pit is maintained through addition of required amount of water. The worms from breeding boxes are introduced in the organic refuse, the worms immediately burrow down into the damp soil.

The compost pit is left for 60 days. It should be shaded from hot sunshine and it must be kept moist. Within 60 days about 10 kg of castinge would have been producer per kg of worms. The pit is then excavated to an extent of about two-thirds to three-quarters and the bulk of the worms removed by hand or by seving. This leaves sufficient worms in the pit for further composting and the pit can be refilled with fresh organic residues and continued.

IN-VITRO VERMI-COMPOSTING

This is also called as bioconversion in soil. This involves the application of the basal dose (5 tonnes/ha) of vermicastings and covering with 2.5 cm. Layer of organic mater (cowdung or pressmud) followed by 10 cm layer of sugarcane trash, crop residues or city swastes. The worms hatch out within 10 days.

VERMI-COMPOSTING OF AGRICULTURAL WASTES

With the aim of vermi-composting of agricultural wastes an experiment was conducted at the Indian Agricultural

Research Institute, New Delhi. The usefulness and efficiency of earthworms in composting of agricultural residues was studied.

Two trials were conducted with mixed organic materials. Trial A, had more of green materials like grasses and Leucanena leaves mixed with soil and paper. Experient B had 4 kg of composting materials consisting of 2 kg of paddy straw, 1 kg soil 500 g twigs, and 500 g shredded paper and was laid in layers per pit. The bottom most layer was laid with twigs to allow for percolation of excess moisture. The compost pit was kept moist and care was taken to avoid waterlogging.

After 10 days of initial decomposition, worms (Eisenia fetida) obtain from M/s Biogenic Ltd., Mumbai were introduced @ 100 worms/pit. Afterthe worms burrowed into the deeper layers, the pit was coveredwith a thin layer of soil.

The results of both of the trials show that introduction of worms in the organic matter piles was found helpful in composting. Organic carbon was reduced at different intervals of composting. In the inoculated compost, nitrogen content wwas appreciably augmented and the C N ratio was narrowed to a desirable level.

The worms were also active in decomposition of high C N ratio of compost was brought to 51 only whereas with worms it was narrowed down to 21. The use of worms also increased the available phosphorous content of compost. Conjoint use of cellulolytic fungi and earthworms showed better result then other of them alone.

TOWN COMPOST

In recent years, large-scale composting of town refuse and night-soil in properly constructed trenches away from human habitations has been taken up successfully by the municipalities of many large and small towns. Trenches, 1 to 1.2 m wide, 75 cm deep and of convenient length, are filled with successive layers of night-soil, town refuse and earth, in this order. The compost gets ready in about three months. The following figures of volume-weight conversion will be found useful in the preperation of town compost.

Volume	**Weight in kg or q**
1 cu.m of refuse	= 318 kg or 3.18 q
1 litre of night-soil	= 0.991kg
1 cu. m of compost	= 636 kg or 6.36 q
1 cartload of refuse(0.849 cu. m),/font>	= 95.40 kg

Social prejudice against the use of this valuable compost has disappeared and town-composting is almost being rapidly adopted in other localities all over India. With a suitable modifications, such as the provision of trench latrines, it can be taken up in villages too. Particulars of the process may may be obtained from the National Extension Service Officers.

SEWAGE AND SLUDGE

The liquid waste, like sulage and sewage contain large quantities of plant nutrients and are used for growing of sugarcane, vegetables and fodder crops near many large towns by operating sewage-farms. In many places, the undiluted sullage has been found to be too strong for healthy plant growth and if it contains readily oxidized organic matter, its use actually reduces nitrates present in the soil. The disadvantages are still greater if sewage is used on land without preliminary treatment. The soil quickly becomes'sewage sick'owing to the mechanical clogging by colloidal matter in the sewage and the development of anaerobic organisms which not only reduce the nitrates already present in the soil but also produce alkalinity. Bacterial contamination makes the eating of raw vegetables on untreated sewage a real danger to health.

For these reasons, it is now usual to construct a setting or a septic tank in which sewage is stand allowed to relieve it of the heavier portion of the solid matter in it, or to undergo a preliminary fermentation. The effluent from the settling-tank, however, still carries a large amount of objectionable colloidal matter, and the deposit of sludge that settles in the tank is of small manurial value and is often offensive. These defects can be removed by thoroughly aerating the sewage in the settling-tank by blowing air through it.

The sludge that settles at the bottom in this process is

called'activated sludge'It has the remarkable property of bringing about the rapid oxidation of the organic matter present in fresh sewage. It is also in offensive and on dry-weight basis, contains 3 to 6 per cent N, about 2 percent P_2O_5 and 1 per cent K_2O in forms that can become readily available when applied to the soil. Similarly, the effluent is a clear, odourless liquid containing nitrates in solution, from shich most of the pathogenic bacteria originally present have been removed. Both the activated slidge and the effluent can be used with safety for manuring and irrigating crops. However, under no circumstances should any produce grown on a sewage-farm be eaten uncooked.

NIGHT-SOIL OR POUDRETTE

Very few towns India are equipped with complete sewage. The sanitory disposal of night-soil with an effective control of foul smell and fly nuisance is, therefore, a serious problem all over the country. Since human excrements are a potential source4 of soil improvement, public health authorities in many towns make the necessary arrangements for its conservation and conversion into a form in which it can be safely used as a manure. The dehydration of night-soil, as such, or after admixture with absorbing materials,e.g. soil, ash. charcoal and sawdust, produces a poudrette that can be easily used as a manure. The mixing of night-soil with an equal volume of ash and 10 per cent powdered charcoal produces an odourless material,containg 1.32 per cent potash and 24.2 per cent time.

The addition of 40 to 50 per cent of sawdust to the night-soil yields straightway a dry, acidic poudrette which may contain 2 or 3 per cent nitrogen. The annual potential quantity of manurial ingredients in the night-soil from India's present population of 600 millions is estimated at appromimately 8.1 million tonnes of dry matter containing almost 0.4 million tonnes of nitrogen, 0.25 million tonnes of phosphoric acid and 0.17 million tonnes of potash.

GREEN MANURES

Despite special efforts at increasing the supplies of

farmyard manure and compost, the supply of farm and other organic manures is scarce and ever becoming more costly. Green-manuring, wherever feasible, is the principal supplementary means of adding organic matter to the soil. It consists in the growing of a quick-growing crop and ploughing it under to incorporate it into the soil.

The green-manure crop supplies organic matter as well as additional nitrogen particularly if it is a legume crop, which has the ability to acquire nitrogen from the air with the help of its root-nodule bacteria. A leguminous crop producing 8 to 25 tonnes of green matter per hectare will add about 60 to 90 kg of nitrogen when ploughed under. Rhis amount would equal an application of three to ten tonnes of farmyard manure on the basis of organic matter and its nitrogen contribution. The green manure crops also exercise a protective action against erosion and leaching.

The crops most commonly used for green-manuring in this country are the following

Sunnhemp (Orotalaria juncea), dhaincha(Sesbania aculeata), cluster-bean(Cyamopsis tetragonoloba), senji(Melilotus parviflora), cowpea(Vigna catjang, V. sinensis), horse-gram(Dolichos biflorus), pillipesara(Phaseolus trilobus), berseem or Egyptian clover(Trifolium alexandrinum). Lentil(Lens esculenta) is recommended in Kashmir for green-manuring paddy.

Sown in late autumn, it is said to provide a winter cover and make new growth in early spring for ploughing under before the sowing of paddy. Sunnhemp is the most outstanding green-manure crop. It is well suited to almost all parts of the country and fits in well with sugarcane, potatoes, garden crops and the second-season paddy in southern India and with irrigated wheat in the north.

Dhainchais in wide use in Assam, Bengal and Tamil Nadu. It does well on alkaline and water-logged soils. Cluster-bean, berseem and senji do well in Punjab, Uttar Pradesh, Rajas than, Delhi and some parts of Madhya Pradesh. Berseem is well suited for orchards and the irrigated crops of cotton and sugacane sown in spring or early summer, as in Punjab and

Uttar Pradesh. Cowpea and horse-gram are used for green locality is naturally the one most suited to its soil and climatic conditions.

Very often, berseem, senji and lucerne (Medicago sativa), and sometimes sunnhemp are grown partly for fodder and partly for fodder and partly for green-manuring. In the case of annual crops of senji and berseem one or more cuttings are taken for use as green-foedder. Lucerne which is allowed to grow for two or three years, is cut seven to eight times for the same purpose. In the case of sunnhemp, the tops are fed to the cattle. In all these instances, the residues (roots and stumps) are incorporated into the soil. These crop residues contain considerable amounts of nitrogen, phosphorus, potassium and other mineral nutrients, besides organic matter. In the case of orchards, the annual green manure-cum-forage crop should be grown at such a time as to interface the least with tree growth and fruit development.

Pulses form an essential part of the India diet, and are grown commonly as pure crops in rotation or mixed with cereals, oilseeds, and fibre crops. Roots and stubble of these pulse legumes return to the soil small quantities of organic matter rich in nitrogen. The inclusion of groundnut on the cotton-jawar rotation of centra, southern and western India, the growing of a quick-maturing variety of ming(Phaseolus aureus) before wheat in Uttar Pradesh, and wheat and rabi jowar in Marathwada Division of Bombay the sowing of peas in the standing crop of irrigated cotton in Uttar Pradesh, and the sowing of sunnhemp in the standing paddy crop in Andhra Pradesh, Tamil Nadu and Karnataka states, or of val IDolichos lablab) in the coastal paddy areas of Maharashtra, are valuable practices for soil improvement.

All these leguminous crops leave the soil in a better physical conditions and richer in nitrogen. The growing of pulses mixed with cereals all over India, the intercropping of cotton with groundnut of with tur(Cajanus indicus) in central and southern parts of the country, or with cluster-bean, mung and moth(Phaseolus aconitifolus) in Punjab and adjoing states; the growing of wheat mixed with peas and gram in northern and central India; and the growing of fodder sorghum mixed with Dolichos lablab in some parts of Tamil Nadu also enrich the soil.

In localities near forests in Tamil Nadu, Karnataka and Andhra Pradesh, the paddy crop is often manured with green forests leaves. These are incorprated into the soil at the time of puddling. In recent years, extensive efforts have been made in these states to plant Glyricidia maculata and Sesbania speciosa on the borders of paddy fields or in other vacant spaces to provide green leaf for manuring the paddy crop. Grown from seedlings or rooted stumps, planted 2 m apart, each Glyricidia plant is said to give annually tow cuttings each of about 6 to 12 kg of green leaf.

It does well in both red black soils. Similarly Sesbania speciosa seedlings planted 10 cm apart on paddy borders produce 1,000 to 2,500 kg of green leaf for manuring 0.4 ha of paddy. Only 115 g of seed is needed to provide seedlings sufficient for border-planting ot two hectares of paddy. In certain other paddy-growing areas, Pongamia pinnata(karanji), Tephrosia, Terminalia and other trees yielding large quantities of leaves are planted for use as a green manure. In the Malabar districts of Tamil Nadu, Indigofera teysmanni is grown to pfovide green leafy twigs for manuring paddy.

For the proper rotting of the green manure, it is necessary that the green material should be succulent and there should be adequate moisture in the soil. Plant at the flowering stage, contain the greatest bulk of succulent organic matter with a low carbon/nitrogen ratio. The incorporation of the green-manure crop into the soil at this stage allows a quick liberation of nitrogen in the available form. With advancing age, the percentage of carbonaceous matter in the plants increases and that of nitrogen decreases. if the material with a side carbon/nitrogen ratio isploughed under, micro-organisms bring about its decomposition, draw upon the released nitrogen and mineral nutrients and cause a temporary nutirient deficiency.

Sometimes methi or dhaincha is sown in-between the rows of the newly planted sugarcane crop, or cluster-bean is planted between the rows of irrigated American cotton. When the leguminous plants are five to six weeks old, they are incorporated into the soil. Sugarcane and cotton are claimed to benefit from the legumes. The decay of green matter in this case seems to occur when it best serves as a fertilizer for the beneficiary crop.

The increase of yield after green-manuring is usually of the order of 30 to 50 per cent. The fertilizing value of the legume crop can be increased a great deal by manuring it with superphosphate. This practice not only increases the phosphorous content of the green-manure plants, but also encourages their plant growth on the whole, thus converting an inorganic fertilizer into an organic manure. Green manures have a marked residual effect also.

FERTILIZERS

Despite the special steps taken in recent years to increase the supply of farmyard and other bulky organic manures, the available quantities are insufficient to meet the existing and prospective needs. Fertilizers have also the dvantage of smaller bulk, the resultant easy transport, relatively quick availability of their plant-food constituents and the possibility of their application in proportions suited to the actual requirements of different crops and soils. Fertilizers are usually classified according to the three principal elements and may, therefore, be included in more than one group.

Nitrogeneous Fertilizers

According to the manner in which their nitrogen is combined with other elements, the nitrogenous fertilizers are divided into four groups; nitrare, ammonia and ammonium salts, chemical compunds containing nitrogen in the amide form, and plant and animal by-products.

Sodium Nitrates

It is also known as'Chilean'nitrate. It owes its importance as a pioneer nitrogenous fertilizer. It occurs in natural deposits in northern Chile and is refined before shipping. The refined product contains about 16 per cent nitrogen in the nitrate form, which renders it directly available tp plants. For this reason, it is highly valued as a source of nitrogen when applied as top-and side-dressings, especially to young plants and garden vegetables, which need readily available nitrogen for quick growth leached out from the soil. for wheat, maize, barley, cotton, sugarcane, etc., it is as benefecial as ammonium sulphate. Sodium nitrate is particularly useful for acidic soils. Its continued and abundant use in soils is said to cause deflocculation and develop a bad physical condition in regions of low rainfall. It should be stored in a dry ware-house

AMMONIUM SULPHATE

It is the most widely used fertilizer in the country. It is a white crystalline salt, containing 20 to 21 percent aoomniacal nitrogen. It is easy to handle and it stores well under dry conditions. During the rainy season, it sometimes, forms lumps. These lumps should be powedered before use. Being soluble in water, it acts quickly, but despite its high solubility, its nitrogen is not readily lost in drainage, because the ammonium ion is retained by the soil particles. it is, therefore, very suitable for wet-land crops, e.g. paddy and jute. It has also been found useful on wheat, cotton, sugarcane, potatoes and many other crops grown on a wide variety of soils. It has, however, an acid effects on the soils. Its long-continued use increases soil acidity and lowers the yield. The application of this fertilizer to acid soils improved the yield of tea plants considerably. It is advisable to use this fertilizer in conjunction with bulky organic manures to safeguard against the ill effects of continued application of ammonium sulphate to field and horticultural crops.

Ammonium sulphate can be applied before sowing, at sowing time, or as a top-dressing to the growing crop. it should not be applied along with, or too close to, the seed, because in concentrated form, it affects seed germination very adversely.

AMMONIUM NITRATE

It is a white crystalline salts, containing 33 to 35 per cent nitrogen, half as nitrate nitrogen and half in the ammonium form. in the ammonium form, it cannot be easily leached from the soil. This fertilizer is quick-acting, but highly hygroscopic and not fit for storage. Granulation of the material and a light coating of the granules with oil reduce hygroscopicity to some extent. It has an acidulating effect on the soil. Under certian conditions, it is explosive, and, therefore, it should be handled cautiously.'Nitro Chalk'is the trade name of a product formed by mixing ammonium nitrate with about 40 per cent lime-stone or dolomite. It is granulated, non-hazardous and less hygroscopic. it contains 20.5 per cent nitrogen, half in the form form of ammonia and half as nitrate. The presence of lime in it makes it particularly useful for acid soils.

AMMONIUM SULPHATE NITRATE

It is a mixture of ammonium nitrate and ammonium sulphate. it is available in a white crystalline form or as dirty-white granules. This fertilizer contains 26 per cent nitrogen, three-fourths of it in the ammoniacal form and the rest(6.5 per cent) as nitrate nitrogen. It is non-explosive and not as deliquescent as ammonium nitrate. It is readily soluble in water and is very quick-acting. its keeping quality is good and it is useful for all crops. Its keeping quality is good and it is useful for all crops. Its acid effect on the soils is only one-half of that of ammonium sulphate. It can be applied before sowing, at sowing time or as a top-dressing, but it should not be applied along the seed.

AMMONIUM CHLORIDE

It is a white crystalline compound, possessing a good physical condition and containing 26 per cent ammoniacal nitrogen. It is extensively used on paddy in Japan, In India, it is used largely in industries. In general, it is similar to ammonium sulphate in action. It is usually not recommended for tomatoes, tobacco and such other crops as may be injured by chlorine.

UREA

It is a white, crystalline, organic chemical. It is a highly concentrated nitrogenous fertilizer, containing 45 to 46 per cent of non-proteined organic nitrogen. it is fairly hygroscopic and presents considerable difficulty, it is also produced in granular or pellet forms and is coated with a non-hygroscopic inert material. It is highly soluble in water and, therefore, subject to rapid leaching. It is, however, quick-acting. When applied to the soil, its nitrogen is rapidly changed into ammonia. Like ammonium nitrate, urea supplies nothing but nitrogen.

It may be applied at sowing time or as a top-dressing, but should not be allowed to come into contact with the seed. It is suitable for most frops and can be applied to all soils.

AMMONIA

It is a gas containing about 80 percent of nitrogen. Under suitable conditions of temperatures and pressure, it becomes liquid(anhydrous ammonia). Another form,'aqueous ammonia', results from the absorption of ammonia gas into water, in which it is soluble. Ammonia is used as a fertilizer in both these forms. Anhydrous ammonia can be applied by introducing it into irrigation water, or directly into the soil from special containers, which makes its use rather expensive. Its possibilities as manure for paddy, sugarcane and cotton are being onvestigated at Bangalore in the Karnataka state. In manurial experiments on cotton in Maharashtra, aqueous ammonia has been found to be as efficient as ammonium sulphate.

Calcium Ammonium Nitrate

Calcium ammonium nitrate is a fine free-flowing, light brown or grey granular fertilzer. It is commercially prepared from ammonium nitrate and ground limestone. It is almost neutral and can be safely applied even to acid soils. Its total nitrogen content may vary from 25 to 28 per cent. Half of this total nitrogenis in the ammoniacal form and half is in nitrate form. According to the prescribed standards, its moisture content should not be more than 0.5% by weight. As regards the particle size, 90 per cent of the material should pass through 4 mm IS sieve and be retained on a 1-mm IS sieve but not more than g per cent shall be below 1 mm.

ORGANIC NITROGENOUS FERTILIZERS

These fertilizers include plant and animal by-products, such as oil cakes, fish manure and dried blood from slaughter houses. Before their organic notrogen can be used by the crops, it is converted through bacterial action into readily usable ammonia-nitrogen and nitrate nitrogen.

These fertilizers are, therefore, relatively slow-acting, but they supply available nitrogen for a longer perios. Furthermore, they may also small amounts of organic stimulants that they may contain, or of some of the minor elements needed by plant.

Oil-cakes of different kinds are produced in India to the tune of about two million tonnes anually. They contain not only nitrogen but also some phosphoric and potash, besides a large quantity of organic matter. In addition to the three fertilizing constituents (N, P_2O_5 and K_2O), the oil-cakes invariably contain2 to 15 per cent of oil, depending on whether the oil is extracted by using solvent process or with expelers, hydraulic presses or indigeneous ghanis. This residual oil, however, dous not, in practice, affect their manurial value. Owing to the great importance of using edible oil-cakes as a cattle feed, their utilization as fertilizers is undesirable. They should be fed to cattle and the excrements may be used as manure.

Inedible cakes, like castor cake, neem, mahua cake and karanj cake can, however, be recommended for use in conjunction with quicker-acting chemical fertilizers. Dried blood or blood-meal contains 10 to 12 per cent highly available nitrogen and 1 to 2 per cent phosphoric acid. It is a very quick-acting manure and effective on all crops and all types of soils. It should be used in the same way as oil cakes. Fish manure is available either as dried fish or as fish-meal or powder. In regions where fish oil is extracted, the residue can be used as a manure. Depending on the type of fish, its manurial constituents vary from 5 to 8 per cent of organic nitrogen and from 4 to 6 per cent of phosphoric acid. It is quick-acting and suitable for all crops and soils. It should preferably be powdered before use.

PHOSPHATE FERTILIZERS

These are classified as natural phosphates, treated or processed phosphates, and by-product phosphates anc chemical phosphates.

Rock Phosphate

It occurs as natural deposits of rock in Morocco, the United States of America, Poland, Russia, Tunisia, Algeria, Algeria, Brazil, Egypt, Nehru and some islands in the Pacific Ocean and the Indian Ocean. It contains 25 to 35 per cent phosphoric acid, but this phosphorus is insoluble in water. Practically no rock phosphate, as such, is used as a fertilizer in this country, except some quantity in southern India. In other countries, finely pulverized phosphate rock has been found to give satisfactory results in soils which are very deficient in phosphprus and are acidic. Adequate rainfall and a long growing period of the crop enhance the response. All the same, very little rock phosphate is used is directly as a fertilizer even in these countries. Much more of it is used to manufacture superphosphate, the phosphoric acid of which is water-soluble and is in an available form.

SUPERPHOSPHATE

It is the most widely used fertilizer in India. Prepared formerly by treating bones with sulphuric acid, it is now manufactured largely by treating ground phosphate rock with almost an equalquantity by weight of sulphuric acid. This treatment produces a brownish-grey mixture containing monocalcium phosphate and calcium sulphate(gypsum) in practically equal quantites. This fertilizer is manufactured in three grades single superphosphate containing 16 to 20 per cent phosphoric acid; dicalcium phosphate, 35 to 38 per cent; and triple superphosphate,44 to 49 per cent. Single superphosphate is the most commonly available grade in the Indian market. Triple superphosphate in the prodiction of which liquid phosphoric acid is used instead of sulphuric acid, contains very little calcium sulphate. Tripple superphosphate is used mostly in the manufacture of concentrated mixed fertilizers.

The phosphoric acid in superphosphate is wholly water-soluble but when applied to the soil, it is immediately converted into insoluble phosphate owing to precipitation as calcium, iron or aluminium phosphate, according as the soil is alkaline or acid. Thus athe fertilizer is not leached, but is slowly dissolved in the soil solution. Fixation losses can be reduced by applying the fertilizer in bands on both sides of the row of seeds at a depth of 10 to 15 cm with a drill. By this means, at least some of the phosphate does not come into direct contact with the soil and thus the available phosphorous is readily released for absorption by the plant roots.

The fertilizer is suitable for all crops and can be applied to all soils. In acid soils, it should properly be used in conjunction with organic manure. It should be applied before or at sowing or transplanting.

Basic-slag

This is a by-product of steel factories. Depending on the phosphorous content of the iron ore, it contains from 6 to 20 per cent of phosphoric acid (P_2O_5). Slag from Indian steel-mills is poor in P_2O_5 and is not used as a fertilizer. The high-grade European slag containing 15 to 18 per cent P_2O_5 is a popular phosphatic fertilizer in central Europe. It is not as soluble as superphosphate, but unlike the latter, it is alkaline in reaction and good for acid soils. For effective use, it must be pulverized before application.

Bone-meal

The use of raw bones as amanure for fruit-trees is an age-old practice in this country. Burying the skeleton of an animal under a fruit-tree is known to benefit its growth and bearing. Large quantities of bones used to be exported until a few years ago. Exports have since declined and bone-meal(i.e. ground bone) is now a widely used phosphate fertilizer. It is available in two forms

- Raw bone-mel,
- Steamed bone-meal.

The steaming of bones under pressure remqves fats, greases, nitrogen and glue-making substances. Thus, while raw bone-meal contains about 4 per cent slow-acting organic nitrogen and 20 to 25 per cent insoluble phosphoric acid.

Steamed bones are more brittle and can be readily ground. This is an advantage, as the rate of availability of phosphoric acid. Steamed bones are more brittle and can be readily ground. This is an advantage, as the rate of availability of phosphoric acid in bones depends largely on their degree of pulverization. Bone-meal contains only 1 to 2 per cent nitrogen but 25 to 30 per cent phosphoric acid.

Steamed bones are more brittle and can be readily ground. This is an advantage, as the rate of availability of phosphoric acid. in bones depends largely on their degree of pulverization. Bone-meal, having particles not larger than3/32 inch, considered suitable for use as a fertilizer, but the more finely powedered it is, the quicker its P_2O_5 becomes available in the soil. Being relatively slow-acting, bone-meal should not be used as a top-dressing; it must be incorporated into the soil in order to become available. It may be applied either at sowing time or a few days before sowing and should be broadcast. It is particularly suitable for acid soils. It is considered a safe manure for all crops.

In some parts of the country, charred and powedered bones are used as a manure. Charring destroys about half the nitrogen, but leaves intact practically the whole of P_2O_5 in a quickly available form. In the absence of arrangements for steaming and grinding, charring can be easily carried out even in remote villages.

POTASSIC FERTILIZERS

Most of the Indian fertilizers soils contain a sufficient amount of potash. Potassic fertilizers should, therefore, be applied only to such soils as are definitely known to be defecient in potash or to those which respond to their application, such as sandy soils. They can also be applied to certain crops, such as tobacco, potato, onion, tomato and fruit-trees, to improve the quality and appearance of their produce. Potassic fertilizers in common use are

- Muriate of potash(potassium chloride),
- Sulphate of potash (potassium sulphate).

These salts are important constituents of the waters of oceans and inland seas and of saline deposits derived thereform. The largest known deposits of these salts are at Stassfurt in Germany, in the Caspian Sea region in Russia, in the Dead Sea in Palestine and at some places in California, New Mexico, france and Spain.

Muriate Of Potash

It is a gray crystalline material containing 50 to 63 per cent potash (K_2O), the whole of which is readily available. Though highly soluble in water, it is not lost from the soil, as it is absorbed on the colloidal surfaces. it can be applied at sowing time or before sowing.

Sulphate Of Potash

It is made by treating potassium chloride with magnesium sulphate and is, therefore, more costly. it contains 48 to 52 per cent K_2O. It dissolves readily in water and becomes available to the crop almost immediately. It can be applied at any time up to sowing, but should not be drilled with the seed. It is considered better than muriate of potash for crops, such as tobacco, chillies, potato and fruit-tree, where quality is of prime importance.

OTHER SOURCES

Wood ashes, cattle-dung ash, leaf-mould, tobacco stems and water hyacinth are the available indigenous sources of potash. Unleached wood ash contains 5 to 6 per cent of potash in the form of potassium carbonate (which is alkaline), 1 to 2 per cent phosphoric acid and 25 to 30 per cent lime (Cao). Both potassium carbonate and lime in the wood ashes counteract the acidity in the soil. Groundnut shell, paddy husk and bagasse ashes are available near decorticating factories and rice and sugar-mills.

These also contain a fair amount of potash and some phosphoric acid. Ground tobacco stems contain 2 to 3 per cent of nitrogen and 6 to 10 per cent of potash in quickly available forms. Hyacinth abounds as a weed in fresh-water ponds in Bengal, Assam, Tripura and Malabar in Kerala. When dry, it contains 1 per cent nitrogen, 4 per cent potassium and a small quantity of phosphorous.

Soil Amendments

Lime is generally used for correcting soil acidity, for improving the physical condition of the soil and encouraging bacterial activity. Similarly, gypsum is used for reclaiming'alkali'soils or land from the sea and improving the structure of heavy black clay soils. Hence, these are called soil amendments.

Compund Fertilizers

These fertilizers are multiple nutrient materials, supplying two or three plant nutrients simultaneously. When both nitrogen and phosphorous are deficient in a soil, a compound fertilizer, e.g. ammophos, can be used. It contains 16 per cent N and 20 per cent P2O5. Its use does away with the necessity of purchasing two different fertilizers and mixing them in correct proportion before use. The nutrient contents of some of the other compound fertilizers are shown below

	N	P_2O_5	K_2O
Monoammonium phosphate	11.0	48.0	—
Diammonium phosphate			
Ammoniated superphosphate	2.0	4.0	
(4 - 7 per cent)	14.0	20.0	—
Potassium nitrate	13.0	—	44.0

Mixed Fertilizers

Compound fertilizers contain plant food elements in fixed propertions and are, therefore, not always best adapted ti different kinds of soils. Accordingly, the needs of different soils can generally be met most economically by the use of fertilizer mixtures containing two or more materials in suitable propertions. Mixtures usually meet nutrient deficiences in a more balanced manner and require less labour to apply than straight fertilizers used seperately. Mixtures containing all the three principal used seperately. Mixtures containing all the three principal nutrients (N, P and K) are termed complete fertilizers.

Some manufacturers prepare special mixtures for different crops, such as wheat, sugarcane, pady, potatoes, tabacco, fruit-trees and vegetables. Trade names like Ammophos, Niciphos and Nitrochalk are employed for other proprietary products, In foreign countries, even insecticides, fungicides and weed-killers, such as DDT, BHC and mercury or copper salts and 2, 4-D are sometimes incorporated into fertilizers mixtures.

When a required mixture is not available of the cost of the mixture sold in the market is high in relation to the cost of its individual components, it may be made at home by mixing the constituent fertilizers in correct proportion. The preparation of mixed fertilizers requires a good knowledge of the properties and mutual reactionof the component straight fertilizers under different climatic and storage conditions. Therefore in preparing such mixtures at the farm or at home, care should be taken to avoid the uneven mixing of incompatible fertilizers, or the mixing which leads to a loss of some of the fertilizing nutrients in the form of gas, converts soluble nutrients into insoluble ones or induces caking. It is unwise to mix the following

- Ammonium sulphate, ammonium chloride, other ammoniacal fertilizers and nitrogenous organic manures with lime.
- Sodium nitrate or potassium nitrate with superphosphate.
- Nitrochalk with superphosphate or lime.
- Ammonium sulphate-nitrate with lime.
- Urea with superphosphate.
- Superphosphate with lime or calcium carbonate or wood ashes.

Ammonium nitrate is an explosive chemical and, therefore, dangerous should be mixing at home. Mixtures containing other nitrates should be made in only such quantities as are to be used immediately, because they absorb water quickly and are not suitable for storage. Bone meal, sulphate of potash and muriate of potash can be mixed with all fertilizers. For further guidance or information on the method of mixing fertilizers, the State Agricultural Chemist or the nearest officer of the Government Agricultural Department should be contacted.

MODE AND TIME OF APPLICATION OF FERTILIZERS AND MANURES

Bulky organic manures should be applied well ahead of sowing, so that the preliminary decomposition takes place before the seeds germinate. Failing presowing application, they may be applied any time after the seedlings have established themselves. They are best applied in the powedered form. A sufficient supply of moisture in the soil is essential for their rapid decomposition. In the case of inorganic fertilizers, potassic and phosphatic fertilizers are best applied just before sowing or transplanting. Nitrogenous fertilizers may be applied either at planting and partly later. Split application is particularly desirable for nitrogen when applied to irrigated crops or to crops in heavy-rainfall areas.

Fertilizers applied before sowing should be broadcast uniformly and harrowed in. In the case of fertilizers containing soluble phosphate, the desirability of applying them in 2.5 to 5 cm wide bands on each side of the row of seeds at a depth of 10 to 15 cm with a drill has already been pointed out. This operation reduces the fixation of soluble phosphate in the soil. It is also a good practice to mix superphosphate with farmyard manure at 18 to 22 kg to a tonne before applying the organic manure, particularly of dairy farms land.

Sulphate of ammonia used as a top-dressing should not be applied when plant leaves are wet. In the case of irrigated crops, the application of fertilizer should invariably be followed by a watering. In the case of fruit-trees, the fertilizer should be applied to the soil under the crown, a few metres away from the trunk. The area of application should be progressively extended as the trees grow bigger.

In advanced countries, fertilizers are usually applied with the help of machinery of diverse kinds and sometimes with an aeroplane or a helicopter. Combined-planters and fertilizer-distributors are employed when row crops are fertilized at sowing time.

DIAGNOSING THE FERTILIZER NEEDS OF SOILS

There are four methods of determining the fertilizer requirements of soil

- Field experiments,
- Pot tests,
- Biological tests,
- Chemical test.

Field experiments contribute the more relaible method, but being time-consuming and expensive, they are conducted mainly by the research farms and research organisations. Farmers wishing to use field experiments as a Valid means of determining the fertility status of their soils should seek the advice of the state agronomist. Improperly conducted field experiments not only mean economic fertilizer practices.

Pot experiments permit test witha alarge number of manurial treatments within a limited space and in a relatively short time. However, as the conditions of such tests are different from those in the field, the results are not always directly applicable to large-scale farming. Biological tests involve the growth of seedlings or of lower forms of plants, such as fungi and bacteria, under specified conditions and the study of their relative growth or the content of needed nutrients. A peridic testing of plant tissues for nitrates and other nutrients indicates the changing needs of crops for different food elements. But these are slow and costly processes and hence not always practicable.

The chemical analysis of soils or of plants growing on them constituents the modern method of determining the fertility status of a soil. Such analysis give information on the relative abundance or scarcity of the different nutrients required by crops from the soil, but they give no indication regarding the exact quantity, of fertilizer that may be applied to make good the deficiency. The dependability of this, method can, however, be increased a great deal by co-ordinating its results with those obtained from field experiments.

Facilities for rapid soil-testing have been made available in almost all states and can be availed of. At the same time, a very large number of fertilizer experiments with different crops on a variety of soils are conducted annually in all parts of the country under a comprehensive scheme. The results of these field experiments, when calibrated against those of rapid soil tests, will make the latter purely dependable. This will then be a valuable aid in the hands of extension workers in furnishing advice to farmers regarding fertilizer practices.

It is also sometimes possible to obtain a clue to the nutrient deficiences of soil with the help of deficiency symptoms in plants, as described already. However, a correct deagnosis of deficiency symptoms needs extensive experience. Furthermore, such symptoms in plants appear long after the actual occurance of nutrient deficiency in the soil. Therefore, such soil deficiencies must be diagnosed and remedied much earlier by adopting other means.

Chapter 9

Chemical Problems

ACIDIFICATION

Acidification has a number of natural and anthropogenic causes. The main natural causes are long term leaching and microbial respiration. The acids found in rainwater (carbonic acid) and in decomposing organic material (humic and fulvic acids) can stimulate leaching by dissociating into H^+ ions and their component anions which then displace or attract base cations from the soil exchange complex. Leaching of bases is most common where precipitation exceeds evapotranspiration, and includes many eastern parts of North America, and northwestern parts of Europe, where soils have been subjected to leaching throughout most of the 10,000-15,000 years of post-glacial time.

Microbial respiration also leads to soil acidification through the production of CO_2 (carbon dioxide), which is dissolved in soil water to form carbonic acid. Other natural processes associated with soil acidification are plant growth and nitrification. During plant growth, nutrient base cations are obtained through root systems in exchange for H^+ ions, thus leading to increased soil acidity. Nitrification is an oxidative process of organic decomposition whereby NH_4^+ (ammonium) ions are converted to NO_3^- (nitrate) ions by nitrifying bacteria, with H^+ ions as a by-product

$$NH_4^+ + 1.5O_2 \Rightarrow NO_3^- + 4H^+$$

These ions are then available for displacing and attracting base cations from the soil exchange complex, thus leading to soil acidification.

The main anthropogenic causes of acidification include certain landuse practices, such as needleleaf afforestation, excessive use of inorganic nitrogen fertilisers, land drainage, and acid deposition resulting from urban and industrial pollution. Needleleaf afforestation has been associated with the acidification of soils and surface waters for a number of reasons. First, needleleaf trees produce litter which is very acidic in comparison with most broadleaf species. Second, because of their high canopy surface area, needleleaf trees are able to 'scavenge' acid pollutants from the atmosphere, later releasing them into the soil via throughfall and stemflow.

Third, due to modifications of the surface and soil hydrology by drainage channels and shallow root networks, water transfer is rapid and is concentrated either at the surface or in the uppermost layers of the soil. Under these conditions, the residence time of the water in the soil is limited, as is the depth to which it can percolate. Thus, the contributions of weathering and ion exchange reactions to the buffering process are limited.

Excessive use of inorganic nitrogen fertilisers in agricultural systems has also been associated with soil acidification, partly through the process of nitrification. If levels of NO_3^- ions in the soil are in excess of plant requirements, they will behave as mobile anions, thus encouraging the leaching process. The acidifying effect of nitrogen fertiliser was demonstrated in a four-year experiment during which different loadings of NH_4NO_3 (ammonium nitrate) fertiliser were applied to grassland plots; soil pH declined markedly with increased fertiliser loading.

Land drainage is another important cause of soil acidification, particularly in soils which contain appreciable quantities of sulphide minerals. Once the land has been drained, soil aeration improves and sulphide minerals are oxidised to form sulphate compounds. Sulphate ions can then combine with H^+ ions in the soil to produce H_2SO_4 (sulphuric acid). Under extreme circumstances, soil acidity may fall below pH 3.0 as in the case of the acid sulphate 'cat-clay' soils which are common in mangrove swamp environments.

Another important anthropogenic source of acidity in soils and surface waters is atmospheric deposition. Here, gases derived from industrial and motor vehicle emissions, particularly SO_2 (sulphur dioxide) and NOx (oxides of nitrogen), are either dissolved in precipitation *(wet deposition)* or deposited directly *(dry deposition)*. Essentially, acidity is derived from H_2SO_4 (sulphuric acid) and HNO_3 (nitric acid) which undergo dissociation in rain and soil water.

Values of pH for acid deposition in industrial areas of Europe and North America are often less than 4.0. Some of the lowest values, however, (pH <3.0) have been observed in acid mists and fogs, a phenomenon known as *occult deposition*. This type of acidity is common in upland areas of Britain where it may be scavenged by the canopies of needleleaf forestry plantations and then transferred to the soil.

The acidification of soils has an impact on a number of soil characteristics. It was report pH reductions of up to 1.5 units over a 55-year period in southern Sweden; these reductions were attributed to high levels of acid deposition derived from other areas of northern and western Europe. Nutrient loss through leaching is another effect of soil acidification. Expert suggests that the supply of available calcium and magnesium in topsoils will be halved within twenty to thirty years, due to continued acidification and leaching.

In addition to nutrient losses, the mobility of aluminium in soils increases as they acidify, especially when soil pH falls below about 5.0. Levels of aluminium are often particularly high in the B (spodic) horizons of podzolic soils. If organic acids predominate in the soil, aluminium is mobilised in the form of soluble organo-metallic complexes. If mineral acids predominate, however, then aluminium is mobilised in its ionic, labile-monomeric form (Al^{3+}); this form of aluminium is particularly toxic to many freshwater organisms, including fish. In a similar way to aluminium, heavy metals, such as lead, zinc and cadmium, are more readily mobilised in acidic soils.

Acidification of soils, and associated nutrient leaching, has also been implicated in damage to trees in forested areas, particularly in central Europe and Scandinavia. It is more likely, however, that nutrient leaching acts synergistically with other factors including ozone pollution, acid deposition, NH_4^+ (ammonium) uptake, drought and frost to produce stress in the tree. Signs of tree damage include needle discolouration, crown defoliation and deformed branch structures. Increased soil acidity and associated changes in aluminium mobility also have serious implications for the quality of surface waters which receive drainage from acidified soils.

Soils vary dramatically in their ability to buffer acidity, a characteristic known as the *buffering* or *acid neutralising capacity*. The buffering capacity of a soil is defined as the amount of acid that needs to be added to cause a reduction in pH of one unit. Soils which contain significant quantities of base-rich, weatherable minerals have a high buffering capacity, whereas those which are dominated by quartz and similarly resistant minerals have a low buffering capacity. In Britain, for example, soils with the highest buffering capacities are found in eastern and southern areas where calcareous parent materials are particularly common. Soils with the lowest buffering capacities are found in areas with base-poor, often siliceous, parent materials which predominate in upland areas in the west and north.

Surface water acidification is widespread in areas where the geology and soils are base-poor, and levels of acid deposition are high. The hydrological pathway taken by drainage waters and their residence time in the soil are particularly important factors in this respect. Water flowing rapidly over the surface or through the uppermost soil horizons undergoes little buffering and may carry organic acids from the soil, therefore contributing significantly to surface water acidification. If water is allowed to percolate slowly through the deeper mineral horizons, however, its acidifying effect is likely to be limited due to increased buffering.

Surface water acidification is exacerbated in areas where needleleaf afforestation is widespread, as is the case in many upland parts of northern and western Britain. Experts examined the chemistry of streams under a needleleaf forest plantation and adjacent moorland in North Wales. The streams under forest were more acidic, and contained higher levels of aluminium, SO_4^{2-} (sulphate) and Cl^- (chloride).

Similarly, in a study of water quality in streams draining land used for rough grazing and forestry, Experts found levels of ionic aluminium to be particularly high in the forest streams. These findings are also consistent with observations in many forested areas. Elevated levels of aluminium and heavy metals in potable water supplies have been implicated in certain mental disorders in humans, notably Alzheimer's disease, although the medical evidence is not conclusive.

In addition to the effects of land use, stream flow conditions have an important influence on surface water acidity. Storm runoff is known to be more acidic than base-flow because it flows rapidly over the surface of the soil, or through the uppermost horizons, and is unable to percolate into the deeper mineral horizons where it might be buffered. Experts report pH values of 4.4 and 6.4 for storm-flow and base-flow respectively in central Wales. Similarly, acid surges associated with storm runoff and snowmelt have been reported in eastern Scotland.

There are a number of approaches to the management and remediation of soil and surface water acidity, including liming, sensible forestry management and reduction of acid emissions into the atmosphere. Liming has long been practised on agricultural land to remedy the problems of acidity, slow organic matter turnover, poor nodulation in some legumes, calcium and molybdenum deficiency, and aluminium and manganese toxicity. Liming materials most commonly used include ground limestone, chalk, marl and basic slag, the main active constituent being $CaCO_3$ (calcium carbonate); other minor components include quicklime (CaO), slaked lime ($Ca(OH)_2$) and $MgCO_3$ (magnesium carbonate).

The lime requirement of a soil varies depending on its buffering capacity and is usually expressed as the amount of $CaCO_3$ (t ha^{-1}) required to raise the pH of the top 15 cm of soil to the desired value. In temperate areas, the ideal soil pH is about 6.5 for arable crops and about 6.0 for grassland. In tropical areas, however, pH values of about 5.5 are often preferred, particularly in soils with high exchangeable aluminium contents, where phosphorus availability may be restricted under more alkaline conditions.

Liming is also used to combat surface water acidity. In many parts of Scandinavia, for example, particularly in Sweden, lakes are often limed directly. This practice is rather costly, however, as the time period over which the lime is effective *(hydraulic retention time)* is relatively short, mostly ranging from several weeks to two years. Furthermore, fish populations do not benefit fully from such treatments as they tend to breed and spawn in the feeder streams which are not limed.

In an attempt to overcome these problems, catchment liming is sometimes practised. Here, the whole catchment benefits from treatment, and the time-scale over which the treatment is effective is usually much greater than when only the lakes are treated. The cost oftreatment may be reduced substantially y strategic applications of lime to specific sites within the catchment, such as spring and seepage areas, and stream headwaters. In some catchments, flow-activated lime wells have been used to supply lime to streams during periods of high flow when acidification is known to be a major problem.

Acidification of surface waters in afforested areas may be controlled through sensible forestry management practices, rather than by chemical means. For example, the termination of drainage ditches before they enter water courses, sometimes in deep check drains which run across the slope, will help to increase the residence time of water in the soil, thus facilitating the buffering process. Other useful practices include not planting in the riparian buffer zones adjacent to water courses, and increasing the proportion of broadleaf trees in plantations.

Most of the acidity mitigation strategies are curative in nature rather than preventative. In the long term, perhaps the most effective strategy is to reduce acid emissions, particularly of SO_2 (sulphur dioxide), into the atmosphere. European emissions of SO_2 from coal- and oil-fired power stations have decreased dramatically since 1980, partly as a result of installing flue-gas desulphurisation equipment and partly by switching to other energy sources such as gas, nuclear power or wind power.

During the mid-1980s, in an attempt to form a more objective and scientific basis to the issue of cuts, and how large they need to be so that long-term ecosystem recovery is ensured, the 'critical loads' approach was devised. The critical load is the highest loading of acid that can be taken by an ecosystem before long-term damage occurs. In some of the most sensitive areas of Britain, it was found that the reduction required for ecosystems to recover is about 90 percent. This is considerably greater than the 60 percent reduction, by the year 2005, to which Britain is currently committed.

Salinisation and Sodification

Salinisation describes the accumulation of salts in the soil, whereas sodification or alkalisation refers to the dominance of the soil exchange complex by Na^+ ions. These processes occur largely in semi-arid and arid environments where evapouration exceeds precipitation, and where soil parent materials and ground waters are rich in sodium salts. In such dry environments, soil water movement is driven largely by evapouration and occurs in an upward direction by capillary action. As water evapourates, salts precipitate out to form saline (solonchak or white alkali) soils. Slight dissolution of these salts and the consequent release of their constituent ions into solution may lead to dominance of the soil exchange complex by Na^+ ions, resulting in the formation of sodic (solonetz) soils.

Saline soils occur most commonly in low-lying floodplain areas where the water table is relatively high, thus leading to intense capillary action. In contrast, sodic soils are found most commonly on slopes immediately above valley floors and flood plains where the water table is lower and drainage potential is greater. On higher slopes in such areas, a leached and more acidic variety of the solonetz soil, known as a solod, is often found. This catenary association of solonchak, solonetz and solod soil types is widely observed in semi-arid areas, including the Indus valley of Pakistan, the Nile valley of Egypt and Sudan, and the Tigris and Euphrates valleys in Iraq and Syria. They are also found in many parts of central and southern Australia and around the Caspian and Aral seas in central Asia. Even in England and Wales, where the climate is relatively wet, salt-affected soils occupy about 6 percent of the agricultural land area, being especially common in low-lying coastal and estuarine areas.

Soil salinity is measured in terms of the electrical conductivity of a saturated extract (ECe), whereas soil sodicity is established from the exchangeable sodium percentage (ESP) or sodium adsorption ratio (SAR). Soils are classified as saline, if the ECe value is 4.0 mmho cm$^{''1}$ or more, pH is 8.5 or less and ESP is less than 15 percent. Sodic soils are non-saline, have a pH of more than 8.5 (usually less than 10.0) and an ESP value of 15 percent or above. Soils are classified as saline-sodic if they are both saline and have an ESP value of 15 percent or above. The salinity of ground waters and other water supplies, which may be used for irrigation purposes, is also assessed on the basis of electrical conductivity (EC). Most water supplies have EC values of 0.15-1.5 mmho cm$^{''1}$, and are said to be of high salinity if the EC value exceeds 0.75 mmho cm^{-}.

Although salinisation and sodification occur naturally in semi-arid and arid environments, they are often exacerbated as a result of human activity. In parts of southwest Australia, for example, removal of indigenous eucalyptus forest has resulted in extensive salinisation and sodification of soils. This has occurred because the deeply rooted trees have been replaced by shallow-rooted grasses and crops, which are less effective at lowering the ground-water level. Capillary action is most intense, and salinity and sodicity are greatest, in soils where the water table is within about 2 m of the surface.

Another important cause of soil salinity and sodicity is poor irrigation practice. Overwatering leads to a rise in the water table which in turn causes enhanced capillary action.

Similarly, poor maintenance of irrigation channels and canals results in leakage of water onto adjacent agricultural land. This has contributed to increased soil salinity and sodicity in parts of the Indus valley in Pakistan, for example.

In the Nile valley in Egypt, river impoundment resulting from construction of the Aswan dam has been associated with a deterioration in soil quality in what were once very fertile floodplain areas. Prior to this, flooding would occur annually in response to heavy seasonal rains in the mountainous source regions of the river. The seasonal flood waters were particularly important to agriculture in the Nile valley for three main reasons—they were an important source of irrigation water, they brought with them fertile silt deposits which were incorporated into the soils, and they helped to flush out salts from the soil. Since river impoundment, the benefits of seasonal flooding have been removed, and the resulting decline in soil quality has been compounded by agricultural intensification, in order to meet both the needs of a rapidly growing population and the demands of overseas markets for cash crops.

Salinisation and sodification of soils are not only associated with arable land, but may also occur in pastoral landuse systems. Experts examined the salt regimes of soils under grazd and enclosed (ungrazed) natural grassland in the Flooding Pampa of Argentina. Here, soils are affected to varying degrees by salts and Na^+ ions, the water table is high and flooding is common during the wet season. The salt concentration in the topsoil of the grazed land increased dramatically and episodically after flooding, whereas on the ungrazed land it did not. This difference was attributed to compaction and impeded drainage, and to reduced vegetation cover on the grazed land, which led to increased evapouration and enhanced capillary action in the soils.

Soil salinity and sodicity have a detrimental effect on both chemical and physical aspects of soil quality. In a chemical context, high salinity and sodicity are associated with elevated soil pH, and under these conditions the availability of certain plant nutrients is reduced, resulting in severe disturbance of the plant nutrient balance as a whole. Elevated salt and Na^+ concentrations in soils are also highly toxic to many plants, although tolerance levels vary between different species. Crops such as barley, cotton and sugar beet, for example, have a relatively high tolerance level, whereas sugar cane, onion and lettuce have a very low tolerance level. As with arable crops, grasses vary in their tolerance to soil salinity and sodicity. Bermuda grass, for example, has a high tolerance level whereas Guinea grass is much less tolerant.

Of particular significance in sodic and saline-sodic soils is the destabilisation and breakdown of soil structure. Deterioration of soil structure is not usually a problem in saline soils, provided that the concentration of soluble salts remains high. However, if salts are depleted by leaching, and Na^+ ions begin to dominate the soil exchange complex, then swelling and deflocculation of clays may occur, thus resulting in aggregate dispersion. These changes, together with translocation of the dispersed clay, may lead to reduced macroporosity and permeability. Although clay swelling may be reversed through increases in salt concentration, deflocculation and translocation are not reversible and may cause permanent deterioration of soil structure.

Soil salinity and sodicity management and remediation are complex issues, and may involve a number of approaches. Appropriate irrigation and drainage techniques are critical in the effective management of saline, saline-sodic and sodic soils. In addition to the amount of water applied, the quality of the irrigation water is important. Use of water which contains high concentrations of salts can lead to increased salinity and sodicity in soils. As the water evapourates, ionic concentrations increase and ions in solution (particularly Na^+) are exchanged with those held on the soil exchange complex (notably Ca^{2+}) until a state of equilibrium is achieved.

Saline-sodic soils in particular require careful management in this respect because, although structural stability is maintained if the concentration of salts in the soil solution is high, if soluble salts are depleted by leaching, structural deterioration will occur as Na^+ ions begin to dominate the soil exchange complex. Reclamation of such soils requires leaching with water of a low enough SAR to facilitate Ca^{2+} exchange for Na^+, but a sufficiently high total salt concentration to maintain soil structure and permeability. As the soil ESP is gradually reduced, the salinity of the leaching water may also be reduced until reclamation is complete. Intermittent leaching is more effective than continuous ponding as it allows time for salts to diffuse into the depleted outer regions of aggregates from where they are removed during the next leaching phase. During continuous ponding, water flows preferentially through the larger inter-aggregate channels and does not remove salts trapped within the aggregates.

Soil salinity and sodicity can also be alleviated using chemical amendments. Sodic soils, for example, can be improved by treatment with gypsum ($CaSO_4.2H_2O$). About 3 tonnes of gypsum are required to reduce by one unit the ESP of a volume of soil which is 1 ha by 15 cm deep. During treatment, Ca^{2+} ions displace Na^+ ions from the soil exchange complex. This benefits the soil in a number of ways—nutrient status is improved, the degree of alkalinity is reduced and structural stability is increased. If $CaCO_3$ (calcium carbonate) is present in the soil, Ca^{2+} ions may be released through the addition of sulphur to the soil. The acidity generated during the oxidation of sulphur to SO_4^{2+} (sulphate) leads to dissolution of the $CaCO_3$; H_2SO_4 (sulphuric acid) may also be used for this purpose.

In addition, salinisation and sodification of soils may be controlled by manipulation of surface microrelief and by reducing evapouration from the soil surface; this helps to reduce the intensity of capillary action. On land which has been irrigated using the flood and furrow method, capillary action may lead to redistribution of salts from the wet furrow areas to the drier ridges where the crops are often planted; to some extent, this problem can be alleviated by modifying the shape of the ridges. Salt concentrations tend to be particularly high on the ridge tops, so planting in this position should be avoided.

Instead, planting should take place on the sides of ridges where salt concentrations are relatively low. Evapouration may be reduced through the addition of organic mulches to the soil surface. Mulching plays a crucial role in the conservation of soil moisture in soils of both warm and dry environments. If levels of soil salinity and sodicity are not too high, it may be possible to utilise salt-tolerant crops and grasses with minimal reductions in yield, thus avoiding the need for ameliorative treatments. The replanting of trees can also reduce the level of the water table and thus the extent of capillary rise of salts towards the surface.

Agrochemical Pollution

In recent decades, the use of inorganic fertilisers has increased dramatically at the expense of more traditional organic nutrient treatments. Between 1950 and 1985, the global use of fertilisers increased from 14 million tonnes to 125 million tonnes, an increase of almost 900 percent. Inorganic fertilisers are used in preference to organic treatments because the nutrients are in a more readily available form and are released rapidly after application. Organic material releases its nutrients slowly, through decomposition processes, and only when conditions are suitable (warm and moist), not necessarily when crops need them.

Fertilisers are applied in a variety of forms—solution, suspension, emulsion and solid. The solid forms vary in particle size from fine powders to coarse granules, and are either spread evenly (broadcast) over the soil surface or mechanically placed, by drilling, into the rhizosphere; generally, the rate of nutrient release decreases with increasing particle size. Fertilisers are based on compounds of plant macronutrients (e.g. nitrogen, phosphorus and potassium) and micronutrients (e.g. zinc, copper, boron and molybdenum), and a variety of nutrient combinations are available depending on the nature of the nutrient problem.

There are five main fates of fertilisers applied to the soil—plant and animal uptake (immobilisation), adsorption and exchange in the soil (fixation), leaching and loss in soluble form through drainage, volatilisation and gaseous losses to the atmosphere (e.g. denitrification), and surface loss in solid form by runoff and erosion.

In terms of plant uptake, the percentage recovery of nutrients varies markedly between the different types of fertiliser. During the first year of application, the recovery of nitrogen from inorganic nitrogen fertilisers is about 50-65 percent, whereas from organic manures it is only about 20-30 percent. Similarly, the recovery of phosphorus and potassium from inorganic fertilisers is about 5-15 percent and 75 percent respectively. In terms of fixation in the soil, many of the phosphorus and potassium fertilisers are of relatively low solubility and the nutrients released are often strongly adsorbed. In contrast, many of the nitrogen fertilisers are highly soluble, nutrient release is rapid, and in most soils adsorption is limited. The differences in degree of immobilisation by plants and animals, and in the extent of fixation in the soil, help to explain why leaching losses of nitrogen are usually far greater than those of phosphorus and potassium. Levels of nitrogen in drainage waters, for example, are often in the range 15-150 kg ha$^{"1}$ a$^{"1}$, whereas levels of phosphorus are generally <1 kg ha$^{"1}$ a$^{"1}$; even in exceptional circumstances in eroded areas, phosphorus levels rarely exceed 10 kg ha$^{"1}$ a$^{"1}$.

Nitrogen losses of up to 30-35 percent of that applied have been recorded for both leaching and denitrification. Leaching is most common in coarse-textured and well drained soils, whereas denitrification losses are greatest in fine-textured, waterlogged and poorly aerated soils.

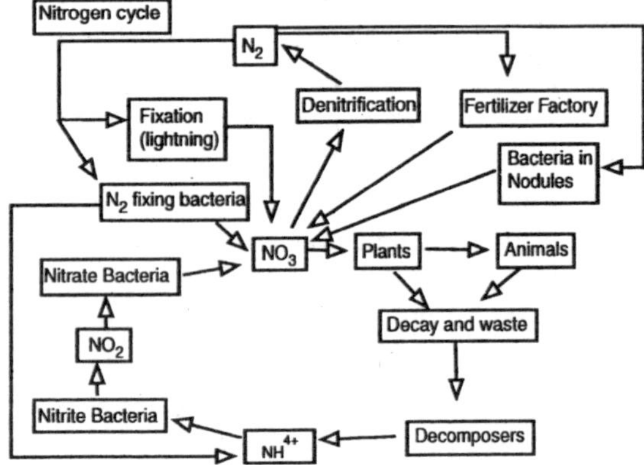

Fig. The soil nitrogen cycle and the fate of nitrogen fertilisers applied to soils

Similarly, surface losses are greatest at sites which are most susceptible to surface runoff and erosion. In terms of the environmental problems associated with fertiliser use, perhaps the area which has received most attention in recent years, by way of research, is that of nitrate leaching.

In the last few decades, levels of nitrate in water supplies have increased dramatically, particularly in intensively cultivated areas where inputs of nitrogen fertiliser have been high. This association is not necessarily causal, however, and may be indirect. Nitrogen fertilisers are not the only source of leached nitrate. It was found that here much of the leached nitrate is derived from the vast reserves of organic nitrogen in the soil, rather than from fertilisers.

The main factors which influence the extent of nitrate leaching include land use, soil characteristics and climate. In terms of land use, nitrate leaching tends to be greatest when there is no crop cover to utilise the nitrate released from fertilisers or organic reserves. With spring planted cereals, for example, the leaching risk is greatest during the wet autumn and winter period when the soil is left bare. In order to maximise uptake and minimise leaching, fertilisers should be applied just before and during the period of maximum crop growth, and large applications at any one time should be avoided.

Tillage practices also have an effect on nitrogen losses through leaching and denitrification. Tillage improves topsoil drainage and aeration and, as a result, rates of organic matter decomposition increase. Expert compared nitrogen losses by leaching and denitrification on conventionally cultivated and direct drilled land in southern England. Leaching losses were considerably greater from the conventionally cultivated land, whereas denitrification losses were smaller. For both tillage regimes, however, maximum nitrogen losses occurred during the autumn period after the harvest, and during the spring following fertiliser application.

Nitrate leaching is not restricted to arable land. Old clover-rich grasslands in particular are able to fix large quantities of nitrogen, and this is released when the grassland is ploughed up. This problem is not as serious in short-term arable leys where nitrogen fixation is considerably less. Even land under needleleaf forestry is not immune to the nitrate leaching problem, although rates of nitrogen release are relatively low in the often acidic and waterlogged soils. Leaching occurs most frequently during the land preparation stages, prior to planting, where improved drainage and aeration lead to increased rates of organic mineralisation, and after felling when the source of nitrogen uptake has been removed.

With respect to soil characteristics, coarse-textured soils are particularly susceptible to leaching due to the often poor structural development and relatively large pore sizes. Rapid drainage in such soils often leads to nitrate contamination of ground-water supplies, particularly in areas which are intensively cultivated. In contrast, clay soils are often characterised by good structural development, small pore sizes and restricted drainage. Under these conditions, surface runoff may occur and leached nitrate is more likely to be transferred into surface waters than into ground waters.

Waterlogging and resulting anaerobic conditions, often found in clay soils, may lead to significant nitrogen losses through denitrification. In soils of tropical environments, which often possess significant anion exchange capacity (AEC), nitrate may be retained in the soil to some extent, thus restricting the leaching process, but this situation does not apply in soils of temperate environments where AEC is usually negligible or non-existent.

In terms of climatic influences on nitrate leaching, rates of mineralisation of organic nitrogen increase with increasing temperature and are particularly high in warm, moist conditions. The timing of individual rainfall events, in relation to mineralisation and fertiliser application, is also important. If a significant amount of rain falls within a month or so of a phase of active mineralisation, or fertiliser application, then nitrate leaching is likely to occur, particularly if there is no crop to take up the released nitrogen, or if uptake is restricted due to poor crop growth. Experts found that levels of nitrate in rivers draining into Lough Neagh, Northern Ireland, were positively associated with nitrogen fertiliser application rates and December to May river flow, and negatively associated with April to September rainfall and June to September air temperature.

Increased levels of nitrate in surface and ground waters have been implicated, together with phosphate, in a number of environmental and health problems. In surface waters, increased nutrient levels lead to excessive algal growth and oxygen depletion, a process known as *eutrophication*. Stagnation, and shading of the bottom environment by algal blooms, causes increased fish mortality and restricted aquatic plant growth. Eutrophication is not restricted to fresh water, but may also occur in marine environments.

In the North Sea, for example, algal blooms have had an adverse effect on salmon and trout farms off the coast of Norway; this damage has incurred costs in excess of $200 million. In spite of this problem, however, countries of the European Union continue to dump 1.5 million tonnes into adjacent marine areas, more than 60 percent of which is derived from agricultural runoff. The nitrate in potable water supplies has important implications for human health.

The incidence of 'blue-baby syndrome' (methaemoglobinaemia), for example, is often greatest where nitrate concentrations in water supplies are high. Similarly, the incidence of disorders of the digestive tract, including stomach cancer, are also associated with elevated nitrate levels, although the medical evidence is by no means conclusive.

In response to these environmental and health issues, legislation has been established in an attempt to control levels of nitrate in potable water supplies. The 1980 European Community 'Drinking Water Directive' (80/778), which actually came into force in 1985, set a limit of 50 mg l^{-1} of nitrate for drinking water; this is equivalent to 11.3 mg l^{-1} of NO_3 -N (nitrate nitrogen). The latest European Union Directive aims to reduce this value by about 50 percent to 5.7 mg l^{-1} NO_3 -N.

In Britain, nitrate levels in drinking water are greatest in intensively cultivated areas where water supplies are derived from deep aquifer sources, particularly in the east and south of the country; these sources contribute about 40 percent of the potable water in England and Wales. In the USA, the Drinking Water Standard sets the legal threshold for NO_3 -N at 44 mg l^{-1}, although this value is commonly exceeded in the more intensively cultivated areas of the mid-west. In Kansas, for example, 25 percent of rural drinking water wells exceed this limit.

Management and remediation of the nitrate leaching problem is a complex issue which depends largely on the adoption of sensible landuse strategies. In terms of nitrogen fertiliser application, slow-release varieties should be used in preference to more soluble types. Similarly, the leaching problem may be reduced if NH_4 -N (ammonium nitrogen) compounds are used instead of NO_3 -N compounds, although increased nitrification may lead to soil acidification. This may be controlled, however, using the specific inhibitor of nitrification, 'N-Serve', in conjunction with the NH_4 -N fertilisers.

It may also be controlled using a mixed fertiliser-lime compound such as 'Nitrochalk'. In landuse regimes where the soil is bare for significant periods of time and the risk of nitrate leaching is high, a 'catch' crop may be grown such as white mustard, forage rape or rye-grass. Such crops take up the excess nitrogen and are then ploughed into the soil just before the next crop is planted; the nitrogen retained by the catch crop then becomes available to the new crop. The creation of riparian buffer zones adjacent to water courses can also contribute to the immobilisation of excess nitrate, thus reducing the risk of surface water contamination.

In a broader landuse context, the 'Nitrate Sensitive Areas' (NSA) and 'Nitrate Advisory Areas' (NAA) schemes, established in central and eastern areas of England where nitrate concentrations in drinking water regularly exceed the European Community 50 mg l″ threshold, provide a useful set of guidelines for management of the nitrate leaching problem. The NSA scheme relies on the willingness of farmers to improve agricultural practices on a voluntary basis, with a view to reducing the amount of nitrate leached from their land.

Guidelines are issued and farmers are encouraged, through incentive payments, to comply with them. The scheme has two levels—basic and premium. The options for these levels, in terms of agricultural practices. The NAA scheme aims to promote good agricultural practice but is not supported financially. Its methods are to avoid exceeding the recommended optimum application levels of fertiliser for each crop, to avoid nitrogen fertiliser applications in the autumn, to avoid large nitrogen fertiliser applications at any one time, to minimise autumn ploughing of grassland, and to avoid application of nitrogen fertiliser to hedge bottoms, field margins and water courses.

In addition to the environmental problems associated with fertiliser application, a number of problems arise from the use of pesticides. A wide range of pesticides has been developed (more than 450 compounds), the types most commonly used being insecticides, fungicides and herbicides; other varieties include nematicides, miticides, rodenticides and molluscicides. Pesticides behave in a variety of ways following application. They may be degraded either biologically or photochemically, adsorbed by organic matter, clay and oxides/hydroxides of iron and aluminium, washed into water courses by leaching (especially compounds with solubilities > 10 mg l^{-1} such as simazine, bromacil and aldicarb) and surface runoff, or they may undergo volatilisation into the atmosphere (especially surface applied compounds, and those of low solubility and vapour pressure such as organochlorines).

Ideally, pesticides should control only the target organism and persist for long enough to achieve this before degrading into harmless products. This is not always the case, however, and a number of environmental concerns arise from pesticide use. These include persistence in the environment, toxicity in soil, vegetation and water supplies, and impact beyond the target organism, including bioaccumulation and its implications for human health.

In fact, the degradation products of some pesticides may be as toxic as the original source chemical but are not often measured by standard assays. The persistence and toxicity of many pesticide compounds and their degradation products are also dependent on a number of soil characteristics, notably clay and organic content, and pH. The environmental impacts of pesticide use were felt most strongly with the early generation of pesticides, particularly the organochlorine compounds and those containing heavy metals. Generally, the organochlorine compounds are the most persistent of the pesticides and may survive for a number of years before they are degraded. Their residues have been found widely in soils, freshwater sediments, fish and cows' milk.

Due to their high lipid solubility, they have also been found in the fatty tissues of animals. All of these findings have important implications for the ecology of food chains and thus for human health. An additional problem was that the effectiveness of organochlorine pesticides decreased as target organisms became resistant to them. The more recently developed organophosphate and carbamate pesticides are much less persistent than the organochlorines, with a half life of about six months, but are often more toxic. The phenoxyacetic acids, 2, 4-D and 2, 4, 5-T, are also rapidly degraded but have been associated with animal and human growth and reproductive abnormalities.

In terms of management and remediation of the environmental problems associated with pesticide use, continued research and monitoring, with the aim of minimising effective persistence and toxicity, and maximising specificity, are essential. Attempts have been made to model the behaviour of pesticides in soils under different management regimes, with a view to reducing the risk of ground and surface water contamination. In addition, the use of granular slow-release pesticides is being explored, together with ultra-low volume application techniques. The environmental problems may also be alleviated by adopting alternative, non-chemical strategies of pest, disease and weed control. These include direct approaches, biological and cultural methods and habitat removal.

Direct approaches are aimed at clearly identified animals and plants, and specific practices include hunting and hand-weeding. Biological methods involve the use of predatory species, preferably with a wide environmental tolerance range, to control the specific target organism. Ideally, as the pest population grows, the predator population should follow, and vice versa; this is known as a *density-dependence relationship*. If such a relationship does not exist, however, the predator itself may become a pest. Cultural methods involve the use of tillage, crop rotation, fertiliser application, liming and drainage techniques. These help to disturb the cycle of pests, diseases and weeds, and to improve the resistance of crops to them.

Habitat removal is perhaps the least acceptable approach to pest, disease and weed control. The basis of this approach is that semi-natural vegetation acts as a refuge for many pests, diseases and weeds, and that control may therefore be achieved through the removal of such habitats. However, this approach fails to recognise both the conservation value of these habitats and the fact that they may be an important source of predators as well as pests. Woodland and hedgerows in lowland arable areas, for example, are an important habitat for ladybirds, essential in the control of greenfly populations, and for bees, important in plant pollination. They also act as shelter belts, thus helping to control soil erosion.

Urban and Industrial Pollution

Urban and industrial development has been associated with both physical degradation and chemical contamination of soils. Problems of physical degradation include erosion, compaction and structural damage resulting from construction activities and opencast mineral extraction. Similarly, chemical problems result from waste disposal activities, discharge and spillage of liquid effluents, and atmospheric emissions, including acid deposition. Soils of urban and industrial environments are every bit as complex and variable in their characteristics as those in rural areas and merit classification in their own right. Experts suggest that the England and Wales soil classification system should be modified to include an 'anthropogenic' major soil group which could be subdivided on the basis of the types of contaminant materials present.

Physical disturbance and chemical contamination of soils in urban and industrial environments are not only recent phenomena. Archaeological investigations have revealed the extensive accumulation of building and domestic waste, although much of this was relatively harmless. Since the Industrial Revolution of the eighteenth and nineteenth centuries, however, the amount and variety of waste materials have increased dramatically. Furthermore, much of this waste is considerably more harmful and often less biodegradable than its early historical counterparts. It is particularly important that any survey of contaminated sites should include an assessment of this historical legacy, since the response of soils to stored pollutants may be delayed for long periods of time. This type of delayed response is potentially very serious and is often referred to as the 'chemical time-bomb' effect.

Scientist identifies four major sources of soil contamination in urban and industrial environments—construction and demolition waste, metalliferous materials, power generation emissions, and chemical and organic wastes. During construction and demolition a range of materials are disposed of in the soil environment, including broken bricks, tiles, glass, timber, piping, wiring and cables, insulation materials, mortar, concrete and plaster.

Once in the soil, these materials undergo a number of chemical changes. Plaster, for example, contains large amounts of gypsum and at sites where the water table is high, the gypsum is dissolved and capillary action may bring it into contact with new concrete structures, thus leading to serious corrosion problems. Similarly, although the use of asbestos and its removal from existing buildings is now carefully controlled, significant quantities can be present in soils and overburden at sites with a long history of construction and demolition.

Metalliferous wastes, particularly heavy metals (e.g. lead, zinc, cadmium, copper and nickel), are commonly found in soils of areas where ore extraction and smelting have occurred. More locally, metal contamination occurs on land used for scrap metal dealing and munitions factories. In an attempt to provide systematic data on the extent of heavy metal contamination in urban soils in Britain, a national survey was commissioned by the Department of the Environment in 1981; 100 household gardens were sampled in each of 53 cities, towns and villages. In most of these localities, total lead levels in the soil exceeded 200 mg kg^{-1}, with values in excess of 500 mg kg^{-1} in industrial areas in the south and in former lead mining areas in the north. These values are considerably greater than the background levels of lead (usually <100 mg $kg^{\prime\prime 1}$) normally found in uncontaminated soils.

Toxic metals may exist in the soil in a number of forms including adsorbed cations, attached to clay and humus colloids, and organo-metallic chelates. Their availability to plants depends on a number of soil characteristics, particularly cation exchange capacity (CEC), pH and the interdependence effects of other metals. In soils with a low CEC, the metals are not retained effectively and are likely to be either leached from the soil or taken up by plants, while in soils with a high CEC, they are likely to be fixed in the soil through adsorption processes. Similarly, the mobility and availability of heavy metals is considerably greater in acidic soils (pH <5.5) than in near neutral or alkaline soils.

Once mobilised, the metals may enter the food chain either through water supplies and aquatic organisms, or through arable produce and grazing animals. The effects of toxic metals in soils on human health are unclear, and it is therefore difficult to establish threshold concentrations above which toxicity problems are likely to occur. However, a number of countries have devised such threshold values, although these vary markedly in response to differences in environmental policy and legislation. The Dutch government, for example, has a system with three levels—an acceptable reference or background value, an indicative value for further investigation and an indicative value for clean-up.

A number of soil contaminants are derived from the power generation industry, including SO_2 (sulphur dioxide) from coal-fired power stations, and radionuclides from nuclear power stations and weapons testing. The anthropogenic radionuclides most commonly found in soils are those of caesium (^{137}Cs and ^{134}Cs). Much of the research in this field has occurred since the accident at Chernobyl in Ukraine in 1986, which resulted in large quantities of radionuclides being deposited across wide areas of Europe, including Britain. The behaviour of radionuclides in soils depends on a number of soil characteristics, particularly clay content and mineralogy, organic content, CEC, pH, NH_4^+ (ammonium) content and nutrient status.

The radionuclides are immobilised most strongly in soils with a high CEC and near neutral pH values, where they are adsorbed onto clays, especially micaceous varieties, and humic materials. They are least well retained in acidic soils with a low CEC, where they may be available for plant uptake. Such conditions are widespread in upland areas of northern and western Britain where the milk and meat of grazing cattle and sheep were contaminated following the Chernobyl accident. In northern Scandinavia, reindeer herds grazing on slow growing lichen heath were similarly affected. The problem of radionuclide entry into the food chain from acid soils in Britain was not anticipated because much of the early research here had focused on clay soils with near neutral pH values where radionuclides are relatively immobile.

A wide range of chemical waste materials have been implicated in the contamination of soils. These include derivatives of detergents, fertilisers and pesticides, paints, dyestuffs, battery chemicals and leather tanning agents, to name but a few. Even silicon chip manufacture involves the use of chemical components, and leakage of solvents used in this process led to the contamination of drinking water in Silicon Valley, California. In recent years, particular concern has arisen regarding the production of dioxins; these are a particularly toxic group of chemicals and are associated with the manufacture of several organic pesticides.

In the Italian town of Seveso, for example, an area of 1,800 ha was affected by dioxin contamination in 1976 and 700 people were evacuated; subsequently, the contaminated soil had to be excavated and removed. A major event which publicised the issue of chemical contamination of soils was the Love Canal disaster near Niagara, USA. Here, extensive dumping of chemicals over a 30-year period led to severe soil contamination by migrating leachates and gasses, with serious health implications for local people.

A number of organic wastes may lead to soil contamination. At Times Beach, Missouri, for example, the land was so badly contaminated by dioxins from waste oils, derived from the manufacture of the defoliant 'agent orange', that it was purchased by the Federal government. Spillage of oils and related wastes is common on land used for storage and maintenance of motor vehicles, and for fuel storage. Sewage sludge, often applied to agricultural land adjacent to urban and industrial areas, often contains high concentrations of heavy metals. The PCBs (polychlorinated biphenyls) are a group of organic solvents commonly implicated in soil contamination. These are used as dielectric fluids in transformers and are often released during the break up of electrical equipment.

Management of contaminated soils is a difficult prospect as the guidelines on threshold toxicity levels of contaminant materials, and on the vulnerability of soils to pollution, are still being developed and refined. Management options depend to a large extent on the geological and hydrological characteristics, and size, of the contaminated area. Such areas vary dramatically from large designated landfill sites with records of waste disposal activities, to former industrial sites with a long and unclear history of pollution. Management options also depend on the nature of the contamination problem and whether it can be contained at the affected site, or whether leachates and gases threaten areas beyond it. Furthermore, it may be possible to treat the contamination on site, or it may be necessary to remove and relocate the contaminated soil.

Experts outline a number of management approaches to the problem of contaminated soil. These include the use of physical stabilisation, barrier, thermal and microbiological techniques. Stabilisation techniques involve treatment to reduce the solubility and mobility of the waste materials. This is achieved through the application of cement, lime, gypsum, silicate materials, epoxy-resins, polyesters or asphalt; these act as binding agents and help to stabilise the 'landform' within which the contaminated waste is stored. Barrier systems usually rely on physical containment using steel or concrete piling to prevent downward and lateral migration of toxic materials.

Similarly, layered cover systems are often used to prevent upward migration of contaminants. Thermal techniques involve heating contaminated soil in rotary kilns or furnaces, whereas microbiological techniques involve the inoculation of waste materials with microbiological communities. Both of these techniques aim to convert the toxic waste materials into less harmful forms. Other approaches to the management of contaminated soil include chemical treatments which may be used to hydrolyse or oxidise contaminants into less dangerous products; similarly, acidic or alkaline wastes may be neutralised. In addition, physical methods may be employed to separate out contaminants according to particle size or density.

The number of sites officially designated as contaminated varies dramatically from country to country. In the UK there are 300 sites, covering an area of about 10,000 ha, where land is officially designated as contaminated. Unofficially, however, there may be between 50,000 and 100,000 sites, covering an area of over 100,000 ha.

Chapter 10

Agricultural Biotechnology

Technological change in agriculture has been the foundation of economic growth and development. By raising yields and continually increasing the number of people who can be fed by the output of one worker in agriculture, change in agricultural technology has permitted an increasing proportion of labour and other productive resources to be devoted to non-agricultural production. Plant and animal breeding and chemical and mechanical technology plus improved husbandry have caused continual structural change in farming.

These technological stimuli have arisen as a series of overlapping waves to create a process of technological change, the momentum of which looks likely to be maintained by intensified adoption of information and biotechnologies. As a result of the cumulative processes to date, farms have become larger and more specialized, and farmers are more highly trained and have substituted machines for animal and human power; farming in the 'West' has become one of the most capital-intensive of industries.

At the same time agriculture has become more dependent upon industry, and even more industrialized-the intensive production of eggs, broiler chickens, horticultural products and pigs has many characteristics which belong to industrial production rather than to traditional farming. Goodman *et al.* talk of 'the industrial appropriation of the rural processes'. Processes of marketing farm products have passed off the farm to be performed by sophisticated distribution and retailing sectors which increasingly dictate details of production to the farm sector, while at the other end of the chain an increasing proportion of inputs is supplied by the chemical, pharmaceutical and engineering industries.

The process of 'industrializing' agriculture has been transferred to the less-developed countries (LDCs) and is epitomized by the so-called 'Green Revolution'. This has entailed expansion of irrigation systems using modern pumps, engineered dams and canals, plus the increased use of inorganic fertilizer, insecticides, herbicides, fungicides and power machinery; although at its heart has been the 'old' biotechnology of breeding new cereal varieties which give high yields when supplied with chemical inputs. Because much of the impetus for this 'revolution' has been western technology and farming know-how, particularly that of the USA, there are those who are critical of what they see as the LDCs' increasing technological dependence upon the West, its industries and research institutions.

What is undeniably the case is that technological advance has been faster, and public and private expenditure on research in western agriculture greater than in LDCs taken as a whole. Thus exportable surpluses from the West have increased with generous public support, while many LDCs have had to resort increasingly to food imports to meet their needs.

While at this stage it is difficult to foresee exactly when, and on what commercial scale, the new agricultural biotechnology 'revolution' which is now brewing will have its effect, most commentators agree that it will intensify the industrialization of agriculture, and that it will increase the technological dependence of most LDCs upon western firms. It may cause considerable disruption to the economies of some countries and will cause farming operations to diverge increasingly from the pattern associated with the country yeoman.

Rural life in many more remote farming areas will be further threatened by the continued deterioration in the economics of ruminant livestock farming on upland and low-productivity pastures. It is these redistributive and structural effects that are the central focus here rather than any attempt to estimate the scale and speed of uptake of specific biotechnology inputs and processes.

Although prospective biotechnological change in agriculture is not different in many fundamental respects from other processes of technological change, in that it will operate through the provision of new or modified inputs (seeds, hormones, etc.) and the creation of new markets in industry, it does raise new issues. Are there dangers from releasing life forms which are engineered using new methods which are quite different from those associated with varieties developed by traditional plant-and animal-breeding methods? Even if the dangers can be objectively assessed and turn out to be minimal, will what may be perceived as man-made and therefore 'unnatural' products be acceptable to society as a whole?

PRINCIPAL DIRECTIONS OF BIOTECHNOLOGICAL DEVELOPMENT

In a recent review of agricultural technology, the Office of Technology Assessment (OTA) of the US Congress defined biotechnology to include 'any technique that uses living organisms or processes to make or modify products, to improve plants or animals or to develop micro-organisms for specific uses-it focuses upon recombinant DNA and cell fusion technologies'. Longworth concurs with this definition, with the significant addition of tissue-culture techniques, an aspect of biotechnology which has great economic, and therefore social and political, potential impact. Despite debates about whether these or any other definitions are adequate, they do touch on the main aspects of biotechnology which deserve to be considered.

The aspect of this new biotechnology which most captures the imagination and stirs the greatest controversy is gene splicing or recombinant DNA techniques, which inspire researchers to consider the possibilities of producing reproducible animals and plants markedly different from current existing species, and referred to as transgenic species. Already in higher animals and plants recombinant DNA has produced transgenic forms which are being commercially exploited. Their agricultural significance is, however, so far limited, and commercial application is concentrated in highly profitable pharmaceutical and horticultural markets. Genes have been introduced into several animal species which alter their protein synthesis to enable transgenic sheep to produce insulin in their milk, and rabbits to produce interferon.

A completely different application has originated in Denmark for salmon, where it has proved possible to introduce germplasm which enables the salmon's physiology to handle heavy metals, which are normally toxic, so opening up new locations for farm fisheries. Much current research is directed to conferring disease immunity on animals, and holds out the prospect of widespread commercial application. In plants one achievement has been the transfer of genetic resistance to antibiotics in the petunia, and another has been the introduction of storage-protein genes from French bean plants into tobacco plants. As yet commercial progress with recombinant-DNA technology in plants appears limited, but extensive opportunities beckon, particularly because plant research is less restricted by the ethical and animal-welfare concerns which apply to research on transgenic animals.

The most important commercial developments based on gene splicing have so far occurred with genetically much simpler microorganisms, and it is with these that the greatest short-to mediumrun commercial potential lies. Already genetically engineered micro-organisms are producing a variety of hormones, vaccines, enzymes and other proteins. Important examples are the production of insulin, vaccines for neo-natal diarrhoea in calves and piglets, and the bovine growth hormone BST identical to that produced in cows which can stimulate a 20 per cent increase in milk yield. BST is already a source of problems for legislators in the European Community, and has provoked strong reactions from the media and milk consumers.

As far as larger animals, and cattle in particular, are concerned, it is developments in embryo transfer and in many processes for manipulating reproduction which hold out the prospect of continuing increases in yields, feed conversion efficiency and general economic efficiency. Already, apparently over 1 per cent of dairy calves in the USA are from embryo transplants, despite the high costs still associated with this procedure. Widespread adoption of these sophisticated technologies would further distance livestock farming from its traditional rural simplicity, and from the natural mating of animals to produce offspring. It would place technical demands upon the operators which favour large-scale company farms capable of supporting a range of highly trained specialists.

Longworth identifies new techniques of tissue and cell culture as having 'the potential for enormous advances in crop improvement in the next couple of decades'. Those techniques 'can both increase the genetic diversity and greatly increase selection efficiency', and they permit innumerable plants to be reproduced asexually from single cells or small pieces of tissue. The particular technique known as 'callus culture' has been used for many years to clone highly valued horticultural plants such as orchids, and cloning of cuttings is widely practised for tree crops such as tea and palm oil as well as by millions of gardeners for garden plants.

According to Longworth 'cell culture' of single cells has unexpectedly, and so far inexplicably, resulted in plants with different properties being regenerated from the same clump of parent tissue. Sugar cane, maize and potato plants regenerated in this way have been found which are resistant to important pathogens. This application of cell culture with the capacity to generate vast numbers of seedlings rapidly has the potential to simplify greatly the hitherto labourious procedures of plant breeding and selection.

However, in terms of current and immediate commercial importance, it is through micro-organisms that biotechnology has its greatest impact. Microbial fermentation processes have been of great commercial significance for centuries, for example in bread, wine and cheese production, and there is the prospect of considerable development. Two recent examples of new processes indicate the sorts of impact that such developments can have. In the late 1960s genetically engineered bacteria were developed which were able to digest corn starch to produce high-fructose corn syrup (HFCS) and which left as a residue corn gluten, which is now an important protein feed for livestock. HFCS has made considerable inroads in the United States and in Japan. It has largely replaced sugar as a sweetener in Coca Cola as well as many other food and drink products.

This has helped depress sugar prices to Third World growers. In the European Community steps have been taken to prevent HFCS and other new powerful sweeteners from being produced and from undermining the market for domestically grown beet sugar, which is supported by the Common Agricultural Policy. A second important application of microbial fermentation has been the production of ethanol from sugar cane in Brazil and from maize in the USA. The commercial viability of these processes is critically dependent upon the price of oil as the main non-renewable source of fuel, and, to date, massive subsidies have been required to maintain the Brazilian and USA ethanol programmes.

Longworth states, however, that a new biotechnology, Sucrotech, is being patented, which will not only reduce the cost of producing ethanol from sugar cane, but will simultaneously produce fructose at a cost which will be competitive with HFCS and may thereby reclaim part of the sweetener market for sugar cane. That such possibilities are in prospect indicates how volatile the future might be; biotechnology has tilted the competitive balance from sugar cane to maize, causing economic pressure and even disruption to sugar-cane-dependent economies and may in future switch it back again.

In the long run the capacity to ferment fuels microbially from renewable agricultural feedstocks points to an important long-term reorientation of agriculture if non-renewable oil becomes uncompetitively expensive as a fuel for cars and feedstock for certain chemicals. It suggests that eventually there will be an increased emphasis on agricultural production of industrial feedstocks at the same time as continually increasing food output will be required to feed the expanding world population. All this will require considerable increases in agricultural productivity, to which biotechnology will increasingly contribute, but it will at the same time impose great strains on the natural environment and the structure of agriculture.

PRIVATE-SECTOR CONTROL OF BIOTECHNOLOGY DEVELOPMENT

It has been the role of the public sector to undertake research and development of agricultural technology as a public service to firms which might profit from translating that R & D into a commercial product or process, to farmers who might profit from adoption of the technology, and perhaps most importantly to consumers at home and abroad who benefited from the lower prices resulting from greater abundance. This was particularly true of the phase of change dominated by improvements in plant and animal breeding, the 'old' biotechnology. While the public sector had a role in basic research for chemical and mechanical technologies, the benefits of research expenditure in these areas were easier to capture by private companies investing in R & D, so that increasingly the public sector has taken a smaller role in these areas although maintaining a strong regulatory role in regard to agricultural chemicals in particular.

Where it is impossible to prevent others from escaping payment for the research costs, either because the product is easily copied (seeds which can be regenerated by farmers or other firms) or because proposals for new methods can be readily implemented, private firms are understandably unwilling to invest. Machines, insecticides, fungicides and other manufactured inputs do lend themselves more readily to private exploitation.

Nevertheless the returns to R & D in these products do depend upon the degree of difficulty potential competitors would have in copying the product. In some cases there are inherent technical difficulties in copying the process, or it would be prohibitively expensive, but in other cases it is the ability to obtain patent protection which creates legal barriers to potential competitors' ability to become 'free-riders' and which protects incentives to private investment in R & D.

Traditionally, however, it has been impossible for plant and animal breeders to obtain patent rights for their products, which is one reason why public-sector R & D has remained so important in this area. For, as under the European Patent Convention (EPC) of 1973, it has been judged impossible for plant varieties to satisfy one of the key criteria to qualify for a patent, namely proof of 'an inventive step'. The application of standard breeding practices to generate new varieties by crossing existing plants or animal strains has not been deemed to be invention. Thus under the EPC one set of exclusions from patentability is 'Plant or animal varieties or essentially biological processes for microbiological processes or the products thereof.'

In the absence of patent rights plant-breeding firms in particular have worked hard to obtain other means of protection and royalties for their products. The history of the development of Plant Breeders' Rights (PER) is presented by Mooney, and is of particular interest because of the concerns he expresses about the consequences of allowing the basic genetic stock, which underpins agriculture and hence our whole society, from becoming private rather than public property. While Mooney's concerns are expressed in relation to 'old' biotechnology varieties of plants, they are of particular importance with respect to the products of the 'new' biotechnology. For bio-engineered products are capable of meeting the inventive-step criterion of patentability, and both micro-organisms and transgenic animals have now been patented in the USA.

Before examining developments in the seed industry it is worth touching upon the implications of changes in research policy which stress increasing reliance upon private R & D to develop and exploit the new biotechnology and other technologies. In the UK government has decided that it should withdraw from funding what it terms 'near-market research', that is, research beyond the basic phase and which is preparation for commercial exploitation. This policy is based in part upon the arguments that industry should be investing more heavily in R & D, and that research strategies at the near-market stage should be driven by assessments of likely commercial success which can best be made by the firms involved, and will therefore lead to greater efficiency in allocating research funds.

This has led to the closure of a number of government-financed agricultural research institutes, the scaling down of others and the sale of the National Seeds Organization by auction to Unilever; Unilever won against competition with BP and other major public companies.

This deliberate attempt to switch an increasing proportion of agricultural R & D expenditure from the public to the private sector carries with it a number of risks. In the first place there is a controversy about whether there is under-investment in R & D so that the returns to extra expenditure are high, or whether the converse is the case. If there is under-investment, then withdrawal of public support for applied research will exacerbate this, as the private sector will only undertake R & D expenditure on those products and processes from which exclusive benefits can be captured by the investor.

Thus in relation to biotechnology in the USA, where again public support for applied biotechnology research is relatively small, Stallman and Schmid (1987) argue that there will be emphasis on technologies which are applied in factory conditions where secrecy and control can be maintained. For technologies which cannot be confined to factories, such as seeds, Stallman and Schmid state that 'Firms are also considering mechanisms which "scramble" the genome of a plant in the second generation', in order to prevent farmers from reproducing seed with the enhanced, engineered characteristics. Clearly such actions are designed to frustrate the maximum spread of benefits from the technology and to maximize private profit for companies investing in research.

Another facet of commercially orientated biotechnology research is that it will aim at the most important crops and livestock products, those produced by the largest farming units, and those which are most heavily subsidized. In the

In order to begin to address ethical issues presented by the genetic modification of foodstuffs, we need to have some ideas of the purposes for which these modifications are performed. (For example, if we wish to approach an analysis on utilitarian cost-benefit grounds.) Despite a great number of general statements, human nutrition and hunger do not appear to be the actual driving forces behind the development of genetically engineered foods.

Other than claiming increased shelf life (which can be indirectly related to nutrition in the sense of reducing spoilage or the consumption of foods that have begun to spoil), there is hardly any indication at all that genetic engineering is being directed to create nutritious substances out of nonnutritious ones; similarly, there are few real instances of genetic engineering reducing the cost of foodstuffs, or increasing the quantities of foods actually available to populations that are hungry because of the non-existence of consumables (as opposed to hungry because they lack money to buy sufficient foods, no matter how produced).

Indeed, early genetic manipulations seemed perversely designed to minimize the achievement of such goals: the creation of 'herbicide tolerant plants' which do not reduce the applications of dangerous chemicals to agricultural fields (which might occur if genetic manipulations of food crops were designed to make them resistant to insects, fungi, or disease) but instead permit higher levels of chemical application, or the introduction of recombinant bovine growth hormone to produce more milk at a time when developed countries suffer from milk gluts and have instituted programmes to kill cows and physically dump milk (milk as a commodity is price inelastic; quantity increases do not lead to price decreases).

The goals of genetically engineering foods seem to be more closely tied to increasing the economic gain and power of the corporations involved in food production (for example, the development of herbicide-tolerant farm crops has been led by corporations which manufacture herbicides), making production easier for corporations, reducing or altering packaging and transportation costs for the corporations, etc.

Environment

Industrial societies around the world are faced by massive pollution legacies from the previous emphases on chemical and nuclear industrial activities. The inability of the US Department of Energy to find a suitable site for a longterm nuclear waste repository illustrates our very poor track record for dealing with the scientific, technical, economic, and socio-political aspects of prior technological 'revolutions'. The environmental problems posed by genetically engineered organisms are likely to be substantially more intractable than those posed by these earlier instances of pollution because genetic wastes multiply, migrate, and mutate. A genetically engineered organism once free in the environment is impossible to recall. (Impacts which are irreversible must be considered with higher scrutiny than those which can be undone.)

Examples of environmental problems which need to be assessed include the risks of transgenic crops themselves becoming weeds; the risk of gene flow to wild relatives which might become weeds or pests; the growth of antibiotic resistance in species (particularly animals), since resistance genes are used as markers for the genetic engineering sites; the problem of exotic or non-native species taking over ecological niches (such as gypsy moth, starling, kudzu, rabbits in Australia, etc.) leading to extraordinary economic losses as well as ecological ones; the restriction (rather than increase) of biodiversity by selective advantageous breeding of transgenic organisms as opposed to natural ones, or the predatory results of expanding exotic species (such as the Dutch Elm micro-organism destroying the American Elm and severely restricting the biodiversity of certain areas in the Northeast United States); health issues (both of plant and animal species, as well as humans — worker health and safety as well as community security); and the occurrence of the completely unexpected, the inability to eliminate uncertainty (for example, the late 1993 floods in the Mississippi valley included the flooding of a field of genetically engineered corn and the dispersal of this plant material to unknown sites within the thousands of square miles of downstream flood plain).

Prospective ecological assessment is not being performed, despite the fact that its need has long since been recognized. And economic priorities, particularly those accruing directly to the promoters, as well as the nebulous spur of 'competition' with foreign countries, oftentimes overwhelms any interest in doing environmental impact analysis.

DEVELOPMENTS IN THE SEED INDUSTRY

Many major food crops upon which we depend originated and were first cultivated in Third World countries rather than in the industrialized countries where they are most productively exploited today and in which they have been improved by traditional plant breeding. The potato originated in South America, wheat in Ethiopia and the Near East, important maize varieties in South America, and rice in Asia. Moreover the so-called Vavilov centres, which contain the greatest variety of living species and which have been the source of much important gene material, are mainly in Third World countries.

The wealth of the industrial countries owes much to their past ability to exploit agriculturally crops and animals originating in other countries. In the colonial era the success of British scientists in smuggling rubber plants from Brazil to Sri Lanka, Singapore and Malaysia and the comparable more complex route by which coffee was introduced to Latin America from Ethiopia are instances of colonial powers obtaining significant wealth by exploiting plant material from the Third World.

In one way or another the process by which institutions and firms in developed countries have continued to collect new plant and animal varieties from less-developed countries has continued. But what is still a major source of friction is the procedures by which legislative protection is developed to give companies in the richer countries commercial rights to exclusive exploitation of varieties developed from plants freely collected from other countries, and to extract high rates of profit from the sale of those varieties. Most extreme is the situation whereby under the USA Plant Patent Act of 1930, which covered asexually propagated plants such as certain fruits, flowers, ornamental shrubs and trees, it has been possible for plants discovered and smuggled out of other countries to be given patent protection in the USA.

The key to this is that the Act accepted the unfeasibility of requiring proof of an 'inventive-step' for the granting of a patent, and employed the criterion that as far as can be determined the plant variety is new. Since some novel variety found in the rain forests of South America may satisfy this criterion, and the source of the plant can be concealed from the authorities, the theft and smuggling of plants can be said to have been encouraged.

Seed companies in developed countries have successfully pressed their case for protection, and have progressively extended Plant Breeders' Rights (PBR) through both national legislation and the Union for the Protection of New Varieties of Plant (UPOV), which is an international agreement among signatory countries to honour a system of varietal rights. In the case of food crops, in the absence of patentability, PBR operates by registering distinct varieties. Pressure has been exerted upon less-developed countries to become signatories of UPOV and thus acknowledge the rights of breeders, most of whom are from developed countries. However, the majority have stalled or rejected membership, and have been fundamentally unhappy about accepting the principle of creating monopoly rights for others over plant material which in some cases originated in their own country.

The whole issue of intellectual property rights has been raised to a prominent political level by the USA's insistence that it be included as one of the thirteen areas for negotiation in the Uruguay Round of multilateral bargaining of the General Agreement on Trade and Tariffs. While the issue is a much broader one than that of plant and animal breeders' rights, it reflects a basic political conflict between the richer countries, determined to obtain secure returns on R & D expenditures by companies, and LDCs struggling to catch up and anxious to avoid having to pay royalties on products protected by PBR and patents. This conflict has broader moral and ethical dimensions when it relates to the issue of charging royalties for the seeds of food crops to countries where malnutrition is widespread.

Against the previous argument has to be balanced the need of those incurring R&D expenditure to capture enough of the returns to provide an incentive for R & D. At least that is certainly the case where R & D is to be undertaken by the private sector, which is politically the increasingly preferred solution in the UK, the USA and elsewhere. This, however, leads to another area of concern, which is that a small number of large multinational chemical and pharmaceutical firms have increasingly come to control the seed industry. Companies such as Shell, Sandoz, Dekallb-Pfizer, Ciba-Geigy and BP have become owners of most of what were small independent seed companies.

This linking between chemicals and seeds within firms has a strong commercial logic. It facilitates the marketing of packages consisting of seed varieties plus the appropri-ate fertilizers, insecticides, etc. It also allows the companies concerned to integrate their plant breeding, biotechnology and chemical research. One particular synergy for these companies will be to bioengineer varieties of major food crops to resist their own herbicides and weedkillers. This would permit chemical weed control to be extended to large-scale crops where this is not now possible, and is just one way in which the direction of biotechnolo-gical research may be influenced (biased) by the combination of forces giving a small number of companies extensive control over the basic means of agricultural production.

There are many other issues which could be touched on regarding corporate influence over the development of agricultural biotechnology. It is, however, understandable that the spokesmen for LDC countries, which are profoundly concerned about their dependence upon western technology and exercised by what is seen as colonial and post-colonial exploitation by the West, should be alarmed by the prospect of new rights being created over seeds and other biotechnological products. Certainly there are widespread fears that they will only gain access to the fruits of biotechnology on unfavourable terms and that their relative dependence on and subservience to western industry will be increased so that benefits to them will be small. There are many western critics who would agree with this, and that the system by which agricultural biotechnology is delivered will favour the 'haves' rather than the 'have-nots'

THE FUTURE OF AGRICULTURAL BIOTECHNOLOGY

There are divergent views about the rate of uptake of new agricultural biotechnologies and their impact. Kalter and Tauer and the US Office of Technology Assessment anticipate comparatively rapid take-up, while others (Buckwell and Moxey and Farrington do not foresee much impact until the next century, that is until ten to fifteen years have elapsed. The caution of the latter commentators is probably justified, for there are various hurdles to be jumped before biotechnologies can gain acceptance.

One hurdle to be overcome is the economic one. Self-evidently there must be some economic advantage to farmers or agro-industry from adopting the technology. Frequently what captures the scientific imagination turns out to be economically unviable in use, although it is only with use that efficiency can be improved and costs brought down.

A more demanding test still for biotechnologies will be to obtain legislative approval and public acceptance. These are intimately connected. In the case of machinery developments, provided safety standards are observed there are no legal impediments to developing new and improved machines, since there are no obvious problems of hazard to the public and therefore public acceptance. Chemical insecticide, fungicides and herbicides do, however, pose much greater problems because of concerns with toxicity, and elabourate testing and licensing procedures have evolved after a relatively haphazard set of procedures in the 1950s and 1960s, when widespread use was made of DDT and organo-phosphorous compounds without adequate recognition of their toxicity.

In significant part, because of the problems which have been experienced with agro-chemicals, the licensing of biotechnologies will inevitably be based on testing at least as stringent as that which now exists for chemicals. Indeed the licensing procedures are likely to be more stringent because of public and scientific concerns.

Public acceptance of biotechnology in the food chain will be made more difficult by confusion about differences between biotechnologies. In the early 1980s there was a crescendo of concern about the use of steroids to promote faster liveweight gain in calves and cattle. The public outcry eventually led to the banning of such hormones for meat animals; but the legacy is a profound distrust of and hostility to new products such as synthetic Bovine Somatrophin (BST), which has the capacity to increase the milk yield of those dairy cows which are 'relatively' deficient in what is a naturally occurring hormone.

Although initial scientific results are favourable, and some minor doubts remain, the major influence leading the European Commission to impose a moratorium on the use of BST until the end of 1990 at least is concern about public perception; the Commission has stated 'It would be a serious setback to producers and to the Community's milk policy were present [positive] trends in consumption to be reversed as a result of adverse consumer reaction.

With newly bioengineered plants and animals scientific and public concerns are emerging which are of a different order from those associated with previous biotechnology. One relates to the possibility that crop failures may become more frequent if biotechnology leads to a reduction in genetic diversity in crops being grown; such a tendency has already been observed in relation to the Green Revolution technology for wheat.

There are also concerns about upsetting the balance of nature in unforseen ways, akin to the unanticipated consequence of introducing rabbits into Australia and then trying to control these by introducing myxomatosis. Thus questions arise such as what happens if herbicide resistance introduced into a commercial crop plant transfers itself by latter case this will worsen the budgetary problems of adjusting agricultural policy in OECD countries. The probable neglect of minor crops, difficult habitats and small farmers means that the public sector will have a defined role in agricultural technology research, but one in which it will be relegated to the second division and where it is unlikely to prove successful in terms of the commercial yardstick of rates of return which is increasingly emphasized by public-research policy.

GENETICALLY ENGINEERED PLANTS AND FOODS

At a symposium at the University of Washington Law School in October 1993 (on the eve of President Clinton's Asian Pacific Economic Council 'Summit'), when discussing the 'Future of Intellectual Property Protection for Biotechnology in the US, EC, and Japan', Joseph Straus of the Max Planck Institute (Germany) warned the participants against 'ethics and other irrational considerations'. Indeed, except for a very narrow band of 'moral dilemma' situations which are the stock-in-trade of professional biomedical ethicists, the ethical aspects of genetic engineering have been routinely ignored by government policy-makers and corporate technology promoters.

My thesis is that it has been the corporate promoters and their governmental handmaidens who have been 'irrational' in their systematic refusal to acknowledge the environmental and ethical considerations of genetic manipulation. Technologies, by definition, are neither acts of God nor nature; they are the embodiment of specific human purposes and intentionality. Far from being inevitable, they are researched and developed by those entities with sufficient power to mobilize social institutions to bring the technology into being, according to their own goals and normative considerations. 'Biotechnology is not neutral. It shares the propensity of modern materialistic science to desacralise, dominate and manipulate life. It reduces all living things to a mechanism which it can manipulate according to engineering standards.'

Genetic engineering is a technological process for undertaking activities which do not, and cannot, occur in nature. In this sense it is perfectly legitimate, therefore, to label genetic engineering as 'unnatural'. The European Community has incorporated this concept in its definition of genetically modified organisms, and it has been adopted by the UN Environment Programm (in the working of its Fourth Expert Panel set up under the Biodiversity Convention as part of the follow through to the United Nations Conference on the Environment and Development held in Rio de Janeiro in June 1992): 'organisms in which the genetic material has been altered in a way that does not occur naturally by mating and/or natural recombination'. Thus, hybrids and other modified organisms obtained by traditional breeding techniques are excluded from the definition of genetically modified organisms.

It is important to note that the process as well as the products are novel. The promoters of the new technology believe that a concern for the process itself is specious, downplaying its novelty whenever confronted by discussions of regulatory oversight. However, one of the bases of the decision of the US Supreme Court in the *Chakrabarti* case (which, 5 to 4, upheld the patentability of genetically engineered microbes, and which has been unquestionably accepted by foreign governments as disposative of the issue of patentability of genetically engineered life forms) was that the element of 'novelty' required by the patent law was to be found in the process by which the microbe had been created.

One reason why we cannot ignore the powerful novel aspects of the processes of genetic manipulation is because we are not omniscient as to what will, in fact, happen when we alter genomes. Genetic manipulation is not like a child's game of Legos or Tinkertoy, in which parts can be rearranged or linked up with only simple mechanical and relatively predictable consequences.

In calculating any risk from a transgenic organism, one should consider four elements: the host organism, the foreign genes, the interaction between the foreign genes and the rest of the genome, and the environment in which the organisms will be used.... [in regard to the last two elements] the literature contains many examples of genetic manipulations where inserted genes did not respond in their new environments the way they did in their old ones or where alterations with one part of the genome caused surprising activity in other parts of the genome.

There is an element of arrogance to scientific assertions which assure us that the process itself poses no risks, since scientists know so little about actual ecosystems; for example, I have been told by agronomists that over 80% of the organisms which can be identified in a soil sample from my garden are completely unknown to the scientific literature. Indeed, the Ecological Society of America itself has specifically warned about the problems presented in our lack of ecological knowledge and the resulting consequences from releasing (intentionally or accidentally) genetically altered organisms into ecological systems.

Foods

In May 1992, the US Food and Drug Administration, responding to corporate pressures to remove the prospect of regulation of genetically altered foods, issued a set of rules which have largely left responsibility for protecting the public health and safety in the hands of the industry, permitting products to be marketed without scrutiny unless the industry indicated to the agency that it believed governmental oversight was justifiable.

However, the proposal as published by the FDA in the Federal Register perversely noted several important problem areas: creating allergens, additions of genes from sources which might violate religious and cultural norms (of vegetarians, Jews or Muslims, etc.), and implications for animal welfare and well-being (for example, the incorporation of human growth hormone-producing gene into a pig's genome produced a highly arthritic animal). Indeed, the commissioner of the FDA, David A. Kessler, and his colleagues noted the possibility that genetically engineered food might 'contain high levels of unexpected, acutely toxic substances'.

cross pollination into a closely related weed species, or indeed if the herbicide-resistant crop should colonize wild habitats. There are, of course, more lurid and absurd notions bandied about in the popular press which are similar in nature to the attacks made on Darwinism.

The questions, both absurd and real, raised in relation to agricultural biotechnology will undoubtedly slow the rate at which it is adopted. Nevertheless instances of adoption are increasing: BST in the USA, a bioengineered baker's yeast in the UK, bioengineered sheep producing insulin, etc. As these become more widespread and increasingly affect lower-valued, bulk agricultural products, so the impacts of biotechnology on structural change will increase. The balance of economic power will switch increasingly to industry, to high-technology large farms and against smaller farmers in disadvantaged regions and countries. That, unfortunately, is a seemingly inevitable consequence of what we consider to be economic progress.

Index

R

S

T

V

W

X

Z